HOW TO SUCCEED AS AN ENGINEER

A Practical Guide to Enhance Your Career

Todd Yuzuriha

PUBLISHING

Vancouver, Washington

J & K Publishing
Post Office Box 87204
Vancouver, WA 98687 U.S.A.

Cover design by Lorna King
This text is printed on acid-free paper.

Publisher's Cataloging in Publication Data

Yuzuriha, Todd

How to Succeed as an Engineer: A Practical Guide to Enhance Your Career.

Includes references.
Includes index.

1. Engineering—Vocational guidance.
2. Engineering—Management.
3. Career development.

Library of Congress Catalog Card Number: 97-94087

ISBN 0-9659084-3-7

To Elaine, who has always believed in me and always been there when I needed her.

To Jill, Ken, and Joy, who are my inspiration and constant source of pride and happiness.

To my parents, who have shown me the value of hard work and perseverance and have always given me the opportunity to pursue my dreams.

And to my in-laws, who have not flinched at even the craziest of my ideas.

ABOUT THE AUTHOR

Todd Yuzuriha is a semiconductor and liquid crystal display expert who has worked over the past sixteen years in all aspects of engineering programs and business development, from product and technology development, to manufacturing, strategic planning, and marketing. Todd is Director of the Corporate Strategic Engineering Center at Sharp Microelectronics Technology in Camas, Washington, managing several multimillion dollar research and development projects. He worked at Tektronix as an engineer and engineering manager before joining Sharp Microelectronics Technology.

Todd is a highly regarded career development instructor of engineers and engineering managers. He is keenly interested in developing the careers of engineers as this leads not only to greater worker satisfaction but also leads to successful organizations. He holds two engineering degrees (a B.S. degree from Stanford University and an M.S. degree from the University of California at Berkeley) and an M.B.A. degree from the University of Portland.

CONTENTS

PART III: FINANCE
97

PART IV: STATISTICS 171

ACKNOWLEDGMENTS

First I thank all the friends, colleagues, and mentors in the fields of engineering, business, and training and development who proofread the manuscript and made suggestions to improve it. They are Dr. Dennis W. Hess, Tom Schmidt, Edwin Cervantes, Ann Fukumoto, James Flores, Dieu Van Dinh, Lisa Y. Johnson, Dr. Martin C. Johnson, Shinya Ichikawa, P.E., Kelli Ambuehl Joy, Dr. Yoshi Ono, Dr. Pradeep Agrawal, Richard S. Hill, Paul K. Fujimoto, P.E., Tai Funatake, P.E., Dr. George M. Homsy, John Hambidge, Robert Okano, and Charles R. White, P.M.P. Their many years of professional experience proved invaluable to make the presentation of this material more meaningful and practical.

I thank Emerson G. Higdon from the Service Corps of Retired Executives for his helpful consultations throughout this project. I thank Amy Lee for her help with library services, and I thank Lorna King for designing the book cover.

And finally, I acknowledge my family. I thank my wife, Elaine, for all her support, understanding, and encouragement in all phases of developing this book. I thank my children, Jill and Ken, for giving up the computer long enough for me to write the manuscript, and now they are writing their own books.

DISCLAIMER

This book is designed to provide information in regard to the subject matter covered. It is sold with the understanding that the publisher and author are not engaged in rendering legal, accounting, statistical, or other professional services. If legal or other expert assistance is required, the services of a competent professional should be sought.

It is not the purpose of this book to reprint all the information that is otherwise available to the author and/or publisher, but to complement, amplify, and supplement other texts. You are urged to read all the available material, learn as much as possible about the engineering field, and to tailor the information to your individual needs. For more information, see the many references listed throughout the book.

An effort has been made to make this book complete and accurate. However, there may be mistakes both typographical and in content. Therefore, this text should be used only as a general guide and not as the ultimate source of information.

In addition, this book refers to many entities as sources of information. Neither the author nor the publisher have personally confirmed the honesty, creditworthiness, or suitability of every entity mentioned. Those who decide to deal with any of these entities as a result of this book should confirm these matters separately themselves.

The purpose of this book is to educate. The author and the publisher shall have neither liability nor responsibility to any person or entity with respect to any loss or damage caused, or alleged to be caused, directly or indirectly by the information or advice contained in this book.

INTRODUCTION

This book is intended to help those of you in the engineering work force and those of you soon to start your engineering career. A practical approach is presented here to enhance your career and achieve greater job satisfaction.

In engineering school, the path to success seems well defined: study hard, get good grades, attract job offers that are the envy of your liberal arts friends, and begin lifelong employment.

Unfortunately once you begin working, the path to success becomes much less clear. Bureaucracy and office politics need to be navigated. You are not necessarily rewarded for successfully completing a project, and promotions often require more than just doing good engineering work. No one in your organization automatically looks out for your well-being. The work which leads to success in school is not always effective in the corporate or public sector. It's common to feel overworked, unappreciated, and underpaid.

How can you survive in today's business environment? How do you excel in today's business environment?

To improve your situation, start with the one thing you have control over—yourself. It is up to you to take charge of your career. Do what you can to improve yourself and maximize your benefit to the company.

In today's competitive environment keeping technically fit is critical. Staying up-to-date with technical advances in your field is imperative to your success. But in order to succeed today, you also need to be armed with tools that extend beyond traditional engineering. These tools include knowledge and practical application of:

- Communication Skills—to know yourself and to work effectively with your manager, co-workers, and customers
- Strategic Planning—to understand the most important tasks to achieve long-term success for the organization
- Finance—to see how you can impact key financial measures
- Statistics—to improve the quality of your technical decisions and your ability to solve problems
- Project Management—to achieve objectives important to the organization while managing resources efficiently.

These tools are not necessarily taught in engineering school, but they are essential for engineers to survive and succeed.

A new level of success can be achieved by engineers who master and apply these tools. This enables you to balance technical mastery with business, quality, and human factors. You must be aware of principles that govern other areas of the organization: Marketing, Finance, Manufacturing, and Quality. With this awareness, you can communicate more effectively with your manager and people from other departments. You are more productive because you have a clearer understanding of the tasks necessary for your organization to succeed and how you can contribute.

The purpose of this book is to show how you can take the initiative and responsibility to improve your career. The practicality of these concepts are presented so they can be applied immediately, if you are not doing so already. These concepts are organized in this book so they are readily available to you, and you can frequently refer back to them. Take charge of your career and take a more active role in helping your organization to succeed in today's business world.

Do not concentrate only on the day-to-day details of your job but try to rise above them to impact both long-term and short-term results. The payoffs for you in expanding your horizons are many. In addition to the enormous satisfaction you derive in contributing to your organization's success, you can also take pleasure in your own personal and professional growth.

PART I

COMMUNICATIONS

Working Effectively With Yourself and Others

Business organizations are becoming increasingly complex and are breeding grounds for communication problems. There are increased competitive pressures. These pressures elevate the importance of teamwork in order to achieve significant results. We all know engineers or engineering students who are extremely skilled technically, but have a hard time communicating their ideas or do not work well with others. There are few jobs which allow a brilliant individual to work in isolation. You must recognize the importance of good communication skills. Communication skills are just as important as technical knowledge for workplace success.

There are three elements at the root of communication problems.

1) **Perception problems**
   Everyone sees the world from their own frame of reference. When you perform experiments, you are trying to gain a better understanding of the world from a technical standpoint. There exists a true state of nature where there are certain mathematical relationships between various elements. You will never have absolute knowledge of the true state of nature, but statistics can be used to design experiments to improve your understanding. Likewise communication skills can be used to gain a better understanding of the true state of nature and improve your perception of situations when interacting with other people. Communication skills help to understand others from their frame of reference.

2) **Lack of listening skills**
   We spend most of our workday communicating. Our communicating time is broken down into writing, reading, speaking, and listening. The education system emphasizes writing and reading and, to a certain extent, speaking. But most of us lack any formal training in listening. Listening is the key to resolving perception problems.

3) **Lack of preparation**
   You increase your chance for a successful career by having a personal strategic plan. This involves setting goals, managing time on a daily basis, and preparing for upcoming communications. Most people lack this type of discipline, but it is essential for establishing a foundation and a direction for your career.

Part I consists of three chapters. The first chapter deals with personal communication—spending time with yourself to set goals and to manage your time. The second chapter deals with interpersonal communication. This will include the principles of good communication, how to listen, and how to present your ideas. The third chapter emphasizes practical implementation of this subject by showing how to get along with your boss and co-workers and how to confront problems when working with others.

Chapter 1

PERSONAL COMMUNICATION

In order to develop good communication skills, start with the only person you have control over—yourself. Take the time to get to know yourself. Bennis says,

> "...until you truly know yourself, strengths and weaknesses, know what you want to do and why you want to do it, you cannot succeed in any but the most superficial sense of the word."[1]

Develop a personal strategic plan. Once this has been defined, you can be more effective at managing your time and will have a foundation to work from when communicating with others.

1. PERSONAL STRATEGIC PLAN

A personal strategic plan is analogous to a strategic plan for a business. As will be presented in Part II, strategic planning forms the backbone of the business. It defines the most important tasks for long-term success. It analyzes the firm's goals, capabilities, and environment.

Similarly, your personal strategic plan forms the backbone of your life. It defines the most important elements to achieve long-term success both at work and outside work. The definition of long-term

---

1. Warren Bennis, *On Becoming a Leader,* Reading, Massachusetts: Addison-Wesley Publishing Company, 1989.

Figure 1.1. Elements of a Personal Strategic Plan

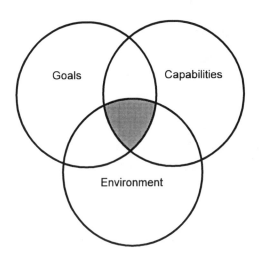

success and how you want to be remembered is up to you. It may include some or all of the following—happiness, family and relationship goals, professional goals, spiritual goals, economic achievements, and recreational goals. As with a strategic plan for the business, your personal strategic plan should analyze your goals, capabilities, and environment (see Figure 1.1).

To begin thinking about your personal strategic plan, use the end of your life as a frame of reference. What are the important goals you want to accomplish by the end of your life? How do you want to be remembered? Consider goals for roles both within your profession and outside your work. Don't neglect health and family. View your life as an integrated whole. As you think of these goals, write them down. Once they are written down, continue to think about them. Share your goals with others. Then rewrite them within 72 hours. Establishing your lifetime goals is the first step toward developing your personal strategic plan.

The next step is to analyze your capabilities. Do you already have the capabilities to achieve your lifetime goals? What capabilities need to be developed? Define your development plan in order to have the skills to achieve your lifetime goals. Again think about goals both inside and outside work.

As you define your development plan, begin breaking your lifetime goals into shorter term goals. These shorter term goals should have durations that are meaningful to you. They can be on the order of one year, six months, or even one month.

Don't forget about professional preparation and personal development. On the professional side, you have started your development plan by reading this book. Study and apply concepts presented in this book in order to make your engineering career more successful. Also consider the education you need to achieve your goals.

Write down your development plan and your shorter term goals. Think about them, and talk to people close to you about them. Revise your development plan within 72 hours.

The final step for defining your personal strategic plan is to consider your environment. Does your current environment allow you to carry out your development plan and achieve your lifetime goals? On the work side, do your professional goals align with the strategic plan for the business? Part II will help you understand strategic planning for organizations. Recognizing how your professional goals tie in with the important goals for the business to achieve long-term success creates a tremendous amount of energy and enthusiasm toward work.

Table 1.1. Steps to Develop a Personal Strategic Plan

Step 1:   Lifetime Goals
Write down the goals you want to accomplish by the end of your life. Consider how you want to be remembered. Once down on paper, think about them, prioritize them, and share them with others. Revise them within 72 hours.

Step 2:   Development Plan
Consider your capabilities in relation to your lifetime goals. List skills that need to be developed in order for you to achieve your lifetime goals.

Step 3:   Shorter Term Goals
Break your lifetime goals into shorter term goals.

Step 4:   Environment
Consider your environment both at work and outside work. Determine if your environment allows you to carry out your development plan and achieve your lifetime goals. If not, you may need to modify your environment or your goals.

Table 1.2. Example of a Personal Strategic Plan

Step 1: Lifetime Goals
1. Raise happy, educated, and well-adjusted children
2. Have a successful marriage
3. Find a way to help engineers with their careers
4. Be an authority in the area of flat panel display technology
5. Maintain a healthy lifestyle
6. Earn enough money to pay for:
    - Retirement from "9 to 5" job by age 65
    - Living and medical expenses
    - Children's college education
    - A house
7. Have close relationships with family, relatives, and friends
8. Speak and read Japanese fluently
9. Play trumpet on a recreational basis

Step 2: Development Plan
1. Spend time each day in activities with spouse and children
2. Work on writing a book to help engineers with their careers
3. Gain technical knowledge in the area of flat panel displays
4. Spend time each day exercising
5. Do not overeat
6. Set aside money from each paycheck for: 401(k) plan, savings account/mutual funds, and home mortgage payment
7. Learn 3 new Japanese characters each day. Study from a Japanese textbook each day.

Step 3: One Year Goals (from 8/16/96 to 8/16/97)
1. Publish a book to help engineers with their careers
2. Give a presentation about flat panel displays at a technical conference
3. Successfully complete the research and development projects as defined at work
4. Lose 10 pounds
5. Save 10% of gross income
6. Be able to read 1000 Japanese characters
7. Be able to carry out a simple conversation in Japanese
8. Play trumpet at a wedding

Step 4: Environment
Current job
- Gives me opportunity to improve technical knowledge about flat panel displays
- Gives me opportunity to improve my Japanese conversation skills
- Requires travel about once a month

Home life
- Have opportunity to work on book
- Have opportunity to exercise, study Japanese, and play trumpet
- However travel takes me away from family--need to pay special attention to my family when I'm not traveling

On the other hand, if your professional goals do not coincide with the business' goals, this may be a signal to consider changing your environment to another job or to adapt to your environment by considering different professional goals. Outside of work, are you in an environment that provides opportunity to achieve your other lifetime goals? If not, you may again have to consider changing your environment or adapt to it.

The steps to develop a personal strategic plan are summarized in Table 1.1, and an example of such a plan is shown in Table 1.2.

By going through this iterative process of analyzing your goals, capabilities, and environment, you should be able to create a first draft of your personal strategic plan. This plan should include your lifetime goals and a development plan consisting of shorter term goals that define how to achieve your lifetime goals. It is important to write this plan on paper because it helps to crystallize your thoughts and strengthen your commitment to achieve your plan. Review the plan frequently. Carry it with you. Update it as necessary. Establishing your personal strategic plan is an essential step toward giving your work and home life a foundation and purpose. It defines your most important tasks to achieve long-term, personal success.

2. TIME MANAGEMENT

Having a personal strategic plan is a prerequisite enabling you to manage your time most effectively. An outline of key time management concepts is shown in Table 1.3.

The first concept deals with effectiveness. Set your priorities based on importance, not urgency. Important tasks contribute toward achieving goals from your personal strategic plan. They require more initiative and proactive thinking. Urgent tasks require immediate attention. You react to urgent matters.

There will always be some urgent matters, but aim to minimize the fire fighting. This can be accomplished with proactive thinking. Try to anticipate problems that may develop. Then prevent them from occurring. An example would be to perform preventive maintenance on equipment. Another example is to spend one-on-one time with those managers or co-workers who may not understand an idea you are proposing or implementing. By taking an active role from the beginning

Table 1.3. Time Management Concepts
1. **Effectiveness**—Set priorities according to importance, not urgency.
2. **Planning**—Discipline yourself to think about and anticipate future events. Force yourself to take time to plan.
3. **Relationships**—Spend time to build and maintain good working relationships with others.
4. **Self-Determined**—Use of your time is set by you, not by circumstances or other people. Make deadlines meaningful and self-imposed.

to solve problems, you minimize the chances of major problems developing such as extensive equipment downtime or managers becoming severely critical of your ideas due to misunderstandings.

Another way to be more effective at managing and using your time is to learn to say no. Tasks that are urgent but not important may be candidates to not do. Obviously, discretion needs to be exercised when saying no to certain tasks. Don't say no unless it is politically correct. If you do not have enough time to handle all your assigned tasks, try to give the person who is asking for certain work to be done a choice of what work could be done. Keep in mind Pareto's law—80% of the results come from 20% of the activities.

Keep focused on your most important tasks. Devote a portion of your time every day toward accomplishing your lifetime goals. Make sure you are doing the right things before you spend time to do things right.

The second concept is planning. Planning is a common theme throughout this book. Planning is essential for a business when developing a strategic plan, and it is essential for yourself when developing a personal strategic plan. It is necessary for setting up control charts and designing successful experiments. Part I shows planning is important for personal and interpersonal communication. In Part V, planning will be shown to be critical for project management.

Force yourself to take the time to plan. As described previously, the first step is to spend time communicating with yourself. Develop your personal strategic plan. Once this is done, take a few minutes each morning to review your schedule. Write down a daily plan based on longer term goals. Be adaptable to changing conditions—the items left over at the end of the day do not necessarily go to the top of the next day's list. Finally, take time to review your personal strategic plan periodically.

The next concept is the human element—make time to build and maintain good working relationships with others. The human element is just as important to your success as technical and financial elements. Chapter 3 advises you how to build constructive relationships with your boss and co-workers.

The fourth concept is realizing you are in control of your time. How you use your time is up to you and not determined by others or your circumstances. You have the freedom to discipline yourself to think longer term, to develop a personal strategic plan. It is within your control to choose to spend time working toward important goals and to anticipate and plan in order to start minimizing urgent items. For instance, if you know you need to write a report about your work every month, plan ahead. Jot down notes for the report throughout the month. Anticipate the deadline, so you can finish ahead of time. If your manager frequently asks for information in an urgent manner, talk to your manager to understand his or her informational needs. Devise systems to automate this information gathering. This allows you to spend less time on urgent tasks.

The last concept is to continually form habits to be more effective at managing your time. Always look to improve. Table 1.4 lists some additional time management tips.

Find out where your time goes by keeping a diary for two weeks. During these two weeks, jot down what you are doing every 30 minutes. Classify your time usage into categories such as time spent fire fighting, time in meetings not meaningful to you, planning time, time for self-improvement activities, and time devoted to building good working relationships. Then decide where your time should go. There should be some time set aside for the important activities like planning and self-improvement that will help you achieve your short-term and lifetime goals. Find out how much time is spent on unplanned, urgent

Table 1.4. Time Management Tips
1. Analyze time usage and make adjustments if necessary.
2. Work on most important tasks during most productive time.

3. Carry a bound note pad or notebook with you to:
    a) Keep your daily plan.
    b) Jot down spontaneous notes and ideas.
    c) Do your thinking on paper.
    d) Listen carefully--take good notes.

4. Have a place for everything--avoid searching.
5. Automate routine tasks.
6. Never lose sight of your personal strategic plan.

tasks compared to planned tasks. Compare your actual time usage to your desired time usage, and make adjustments if necessary.

Determine your most productive time. Make sure you are working on your most important tasks during that time. Some people create uninterrupted blocks of productive time by going to work early or staying late.

Carry a bound notebook with you to organize your thoughts, notes, and ideas. Use this notebook to keep your daily plan and future plans. Write down your commitments as you make them. Take good notes of conversations and meetings attended in this notebook.

Develop a filing system where you have a place for everything. Being organized and able to access information quickly shows professionalism and minimizes time wasted looking for things. Store information you need immediate access to in your bound notebook. For other items, try filing them by subject rather than by chronological order. For large topics, use notebooks pertaining only to that subject. For smaller topics, file them alphabetically in a file folder system, tabbed notebooks, or an electronic mail folder system. Keep well-organized subdirectories in your computer.

Automate routine tasks where possible. Computers are often essential to automate tasks. Take the time to write programs, subroutines, and macros for repetitive tasks that will save you time in the long run. Use common report templates within your word processor and customized address lists for your electronic mail. For repetitive reports that you write, contact the recipients to make sure the information you provide is useful. If it is, try to automate the data gathering process so more time can be spent analyzing and interpreting the data.

Finally, never lose sight of your personal strategic plan. Carry it with you, review it, study it, and modify it as necessary. Your plan links together your activities at work with important activities outside of work and serves as a foundation from which to live your life.

Chapter 2

INTERPERSONAL COMMUNICATION

The last chapter showed that in order to improve your communication skills, start communicating with yourself. Spend time planning, managing your time, and building constructive relationships with others. This chapter points out how to build these relationships and how to communicate effectively with others.

No one can deny being able to work with others is essential for optimum performance. You need to be able to interface with a number of different people on a variety of levels. However when interfacing with people at any level, there are certain principles that will make you more successful. These principles are described below.

1. PRINCIPLES

Table 2.1 lists principles for good interpersonal communication.

1.1. Seek To Understand and To Be Understood

The introduction to Part I pointed out no one has perfectly correct knowledge of the true state of nature. We all form our own opinions about circumstances that arise based on how we perceive the world, filtered through our beliefs. This is the case when interacting with other people as we carry out our professional and personal lives. The first principle of good interpersonal communication states the most effective communication is two-way communication. Seek both to understand others and to be understood. The key to understanding others is be a good listener. The key to being understood is get your

Table 2.1. Principles for Good Interpersonal Communication
1. Seek not only to be understood, but to understand.
2. Build trust with others.
3. Respect the feelings of others.
4. Examine purpose and clarify ideas before each communication.
5. Keep an open mind to debatable points.
6. Be mindful of overtones as well as basic content of message.
7. Emphasize face-to-face communication.
8. Time communications carefully.

ideas across. Tips on how to accomplish these will be described later in this chapter.

1.2. Build Trust with Others

Building trust with others is essential for good interpersonal communication. When you have a high level of trust with others, communications are open, honest, and easy. You can even make mistakes without the relationship suffering. You have more flexibility on the job. Trust brings out the very best in people.

When you have a low level of trust, communications are strained. You must always keep your guard up and be careful what you say. In a low trust environment, there are usually a flurry of memos when an issue is being "worked out". You lose freedom and flexibility because you feel resistance to taking any action or making any decision on the job.

Ways to build trust with others include delivering on your commitments faithfully. Make sure your actions reinforce your word. Be honest and kind.

Ways to lower trust with others are not keeping your promises, being disrespectful, overreacting, and cutting off the other person.

1.3. Respect Feelings of Others

Respecting the feelings of others ties in with building trust and understanding. It is important for maintaining the self-esteem of the people with whom you work. Listen to the viewpoint of others; try seeing from their perspective. The most effective interpersonal communication occurs when there is a good working relationship between you and the other person. Be aware of communications that achieve short-term gains at the expense of damaging long-term relationships. An example is making a decision that significantly impacts the work of somebody else. There may be time pressure, so you make the decision and announce it publicly without first consulting privately with the other person. The short-term gain is you take action quickly, but the long-term loss is a strained relationship. Sometimes this type of communication is necessary, but too many of these communications will destroy a relationship. Communicate for tomorrow as well as today.

1.4. Examine Purpose and Clarify Ideas

This principle is part of a major theme throughout the book—planning. Force yourself to take the time to examine the purpose and clarify your ideas before each communication. Where appropriate, consult with others when planning. Ask for advice and viewpoints of others to make your communication more effective. The clearer you can be in presenting your idea and having the necessary supporting information, the more effective you will be. Details for making successful presentations are discussed later in this chapter.

1.5. Keep an Open Mind

As you try to understand the viewpoint of others, keep an open mind to debatable points. Be receptive to the ideas of others. Say less than you think as you keep your ears open. On points you do not agree with, discuss them but do not argue. The sign of a good working relationship and a mature mind is to disagree but remain friendly.

1.6. Be Mindful of Overtones

As you try to understand the message of others, pay attention to overtones as well as the basic content of the message. Be aware of the presenter's tone of voice and body language. When talking with others, only 10% of the communication is represented by the words we say, 30% is communicated by our sounds and inflections, and 60% is communicated by body language.[2] Therefore how you say something is often more important than what you say. Many people have difficulty with voicing their disapproval of an idea or rejecting a request. They may try to communicate in more subtle ways. For instance, signs the other person may have a negative opinion include not making eye contact with you, crossed arms, or their body is turned away from you when they talk. Be mindful to the possibility of hidden meanings that are not discussed directly.

1.7. Emphasize Face-To-Face Communication

In this age of high technology, there are increasingly new ways to communicate—e-mail, voice mail, faxes, cellular telephone, and televideo conferencing to name a few. However what still rings true is the following principle: emphasize face-to-face communication. The more important the communication and the more important the message be understood, the more critical it is to communicate face-to-face.

Ineffective communicators tend to rely too much on new technology communication methods. New technologies are difficult media to use to convey points about controversial issues. Consider how many times you have been caught in the cross fire of memo wars with e-mail or faxes. Emotions escalate as the discussions progress. E-mail and faxes are too impersonal for communicating disagreements and criticisms. They can lead to strained working relationships when used to settle controversial issues.

The term flaming refers to sending a negative emotional outburst (such as anger) in an e-mail message. Avoid flaming when sending e-mail. If you receive a flaming e-mail message, avoid sending a flaming response. Instead, talk to the other person face-to-face if at all possible.

Surely the new technology methods are great for facilitating routine communications such as setting up appointments, keeping people updated on routine projects, and keeping in touch with people. Using

---

2.   Stephen R. Covey, *The 7 Habits of Highly Effective People,* New York: Simon & Schuster, 1989.

new technology increases your efficiency and saves time with routine communications.

However for effective communications concerning non-routine topics, start with face-to-face, one-on-one communication. In this way you can get real time feedback and can minimize any misunderstandings from the start. Talking individually with the other person gives you the opportunity to pick up any non-verbal cues that is not possible with other methods. You can listen to the concerns of the other person and respond interactively. Face-to-face, one-on-one communication is still the most effective two-way communication tool.

1.8. Time Communications Carefully

As the pace of the work environment accelerates, there are more projects, tasks, and people to keep in touch with. Consequently it is more difficult to get the attention of the other person with whom you want to talk. That is why it is important to time your communications carefully.

Know your audience. When planning face-to-face communications, find out the best time when the listener can devote as much attention as possible to the subject. If you don't know when the best time is, ask the other person or the person's secretary. Set up an appointment for your communication meeting. The more temperamental the person, the more important timing is to be effective.

A couple of themes forming the foundation for good interpersonal communication are knowing how to listen and how to present your ideas. These topics will now be covered in greater detail.

2. HOW TO LISTEN

As pointed out in the introduction to Part I, a lack of listening skills is one of the main causes of communication problems. The education system emphasizes reading and writing. There is hardly any training for listening.

The goal when listening is to understand. Better understanding enhances your working relationship with the other person. Taking the time to listen improves the self-esteem of the individual. Better understanding greatly improves your problem solving ability. Here are some tips on how to listen.

2.1. Stop Talking

To begin listening, you need to stop talking. This is an obvious point, but it is amazing how often this point is forgotten. Have you ever heard the expression "verbal diarrhea"? If you think you are talking too much, you generally are talking too much. Try to put skid chains on your tongue and open your ears.

2.2. Utilize Mental Spare Time

One of the problems while listening is your mind has some free time. People speak at 125 to 250 words per minute, but you can listen at 500 words per minute.[3] Thus your mind can wander for more than 50% of the time.

Good listeners utilize this mental spare time by thinking about what is being said. Listen for ideas and concepts. Stay focused on the meaning and not on appearance or delivery style. Evaluate the development of the idea by thinking back on what has been said, and try to anticipate what the speaker is going to say. Go through the thought processes of the speaker, but avoid jumping to conclusions before all the facts have been presented.

2.3. Be Aware of Emotional Filters

To reiterate, the goal of listening is to understand. However many times we don't practice this. Instead we listen with the intent to reply. We filter what we hear through our own frame of reference, our paradigms. Then our understanding of the message gets clouded by our attitudes and biases.

In order to truly understand the other person, try to see the situation from the other person's frame of reference. Perceive the message in terms of its intellectual content and its emotional content. Adopt the other person's feelings as if they were your own. Empathize with the other person. You do not need to agree with the other person. Again your main purpose in listening is to hear and understand the message being sent.

---

3.   David A. Whetten, and Kim S. Cameron, *Developing Management Skills,* Glenview, Illinois: Scott, Foresman and Company, 1984.

2.4. Ask Questions

Ask questions about what is said in order to improve your understanding. As a listener, you should be skilled with the use of open-ended and closed-ended questions to help guide conversations to maximize your understanding.

Use open-ended questions to begin or expand the discussion. Examples of open-ended questions are:

"Please tell me more about...?"

"How do you see the situation...?"

When asked open-ended questions, the speaker should become more willing to discuss his or her views.

Use closed-ended questions to ask for specific information. Closed-ended questions are those that can be answered by "yes" or "no". Examples of closed-ended questions are:

"Did you observe this (specific situation)?"

"Can this (specific action) be done?"

These type of questions can be very useful to find out certain pieces of information that have not been covered by the speaker.

As the conversation progresses, take the time every once in a while to check your understanding. Restate in your own words what you have understood. Try not to use the exact same words as the speaker—try to paraphrase.

2.5. Summarize the Discussion

At the end of your conversation, close by doing the following.

- Summarize what you understand of the discussion's key points. Check with the speaker to make sure there are not any misunderstandings. With this last check, you can make sure the main items were not misinterpreted.

- Agree on what the next actions will be. This includes deciding on an action plan and scheduling time for a follow-up discussion. The action plan should include who will do what by when in order to address issues or concerns which surfaced during the conversation.

- Always show appreciation for the time and effort taken on the part of the speaker in order to help your understanding. This is an important step to enhance your working relationship with the other person.

Listening to understand is very time-consuming. It is difficult to find the time to have meaningful conversations with others given the day-to-day pressures at work. But the time you take in the short-term to truly understand others will pay dividends in the long-term. The payoff can be seen in terms of improved interpersonal communications, improved interpersonal relationships, and more meaningful results.

3. SETTING UP FEEDBACK SYSTEMS

Engineers know about feedback. In order to ensure an electrical, mechanical, physical, or chemical process is performing with optimal effectiveness and efficiency, feedback is necessary. A typical, simple engineering system is shown in Figure 2.1(a). There is an input and an output. Feedback is taken from the output to monitor how the process is going and to make corrections if necessary.

Similarly when working with others, feedback is necessary for optimal effectiveness and efficiency. The people to whom you supply a product, information, or a service are your customers. These customers can be internal or external to your work organization. The traditional view of a customer is they include only external customers—those people who buy your goods or services. However you must also recognize your internal customers. Your manager, people to whom you write reports, and departments who rely on your technical creations are examples of internal customers. A typical, simple customer-supplier system is shown in Figure 2.1(b). There is a supplier and a customer. Feedback is taken from the customer in order to monitor how the process is going and to make corrections as needed.

Setting up feedback systems with customers is one of the most important tools for getting results. There is no doubt you need to be competent with the technical aspects of the job. However to be the most successful, you also need to respond to the financial and human elements of the business. Key measures of the financial accounting system will be covered in Part III. Just as important is the human accounting system which requires setting up feedback systems. Use the feedback obtained from customers to help guide your actions.

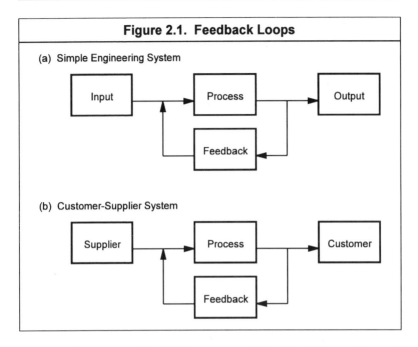

Figure 2.1. Feedback Loops

In order to set up feedback systems, utilize the skills you learned earlier on how to listen and the principles of good interpersonal communication. First identify your customers. List those people inside and outside your organization that depend on you for various goods, services, projects, or information. Next list the product you supply by each customer's name and why you think it may be important to them. Set up regular intervals to gather information from these customers. The intervals depend on the frequency and importance of your interactions with each customer. The more frequent and important your interactions are, the more frequent your intervals for obtaining feedback should be. These intervals may range from once a week to once a year.

The method employed for obtaining feedback is up to you. The most effective method is one-on-one, face-to-face communication. However this is the most time consuming and may not be practical for all your customers. Sending out a questionnaire form is the most efficient in terms of time invested to obtain the information, but the quality of information may not be the best. It is also possible to employ some combination of one-on-one communication and questionnaire forms.

The main information to extract from the customer is their current level of satisfaction with you as a supplier and how the customer's satisfaction can be increased. Listen with the intent to understand, and do not become defensive. Put yourself in the shoes of the customer. Analyze the data you obtain. Compare it with previous data from the same customer as well as from other customers. Is the customer's satisfaction improving? What items have been the most successful and effective for improving customer satisfaction? Identify specific plans and action items that will improve your customer's satisfaction.

The time taken to set up and utilize customer information will pay off many fold in the long run. As you gather and analyze your customer satisfaction information, you will become better at pleasing the customer. This enables you to improve your interpersonal relationships with customers, and makes you more effective in dealing with people and handling people problems.

4. PRESENTING YOUR IDEAS

Like it or not, much of your effectiveness as an engineer is measured by your ability to speak. Being able to make clear, concise, and compelling presentations has a significant impact on your career development. The purpose of this section is to show you how to make a presentation. Whether it's for an audience of one or a thousand, the key words to remember are focus and preparation.

4.1. Focus

Always stay focused on the main points of your talk. Everything in your presentation should either state, explain, or develop your main points. Any other information should not be used for your presentation.

The first step is to identify the main point or points. Is the purpose of your presentation to communicate information or something more controversial? Write down the purpose and main points on a piece of paper. Use Table 2.2 as a guide.

Your presentation consists of three parts: the beginning, middle, and end. Make your main points a major focus for each part. Audiences are getting less patient, so go right to your bottom line message from the outset. In the beginning, state the purpose and main point of your message. Use verbiage such as:

Table 2.2. Presentation Preparation Form

Purpose	Main Points

Audience Profile

Who	Interests and Background

Likely Questions	Answers

"The purpose of my talk today is..."

"Today I propose the following idea..."

"In my talk I want to discuss..."

In the middle, develop your ideas in a logically compelling way. The development of your main points should be as clear and concise as possible. At the end, conclude by stating your purpose and main points once more. Basically, tell the audience what you will tell them (beginning), tell them (middle), and then tell them what you have told them (end).

Studies estimate listeners forget 40% a half hour after the presentation, 60% by the end of the day, and 90% by the end of the week.[4] What remains in the listeners' minds are impressions. Therefore, it is vitally important to keep your talk simple and to repeat your main points in the beginning, middle, and end. Create as many visual images as possible. These tactics maximize the chance your audience will retain your message.

4.2. Preparation

To prepare for your presentation, jot down your answers to the sections in Table 2.2. From the previous segment, you have already written down the purpose and main points of your talk. To help use this form, Table 2.3 shows an example. This example is for the case of selling your idea to utilize Design of Experiments (DOE) to speed up the introduction of new products. The purpose and main points are listed in Table 2.3.

Next, profile your audience. Anticipate who will attend your presentation and write down what you think their special interests and hidden agendas are. Jot down the personalities of the key people. Center your talk around the audience and as much as possible, tell them what they want to hear. Determine the viewpoints of key personnel and how to reach them. Usually, individuals are only interested in a presentation to the extent they know what is in it for themselves.

---

4. Karen Berg, and Andrew Gilman, *Get to the Point: How to Say What You Mean and Get What You Want,* New York: Bantam Audio Publishing, 1989.

Table 2.3. Example of Using Presentation Preparation Form

Purpose	Main Points
To convince the engineering group to use Design of Experiments (DOE) when characterizing new products before introducing them to market. This includes getting approval for DOE training ($50,000).	1. Get to market 2 months earlier--DOE will reduce the characterization time from 3 months down to 1 month. 2. Lower costs--DOE will reduce the number of samples needed for characterization. 3. Effects of interactions between variables can be quantified.

Audience Profile

Who	Interests and Background
1. Engineering Director	(a) Meticulous with technical details (b) May be defensive because he didn't come up with this idea. (c) Wants to look good for his boss
2. Quality Assurance Manager	(a) Good statistics and DOE background (b) Familiar with DOE benefits (c) Should be helpful with program implementation
3. Engineering Group	(a) As a whole, not familiar with statistics or DOE (b) May question how more information can be obtained with less samples

In the example shown in Table 2.3, the Engineering Director is picky with the technical details. The Q.A. Manager has a good background in Design of Experiments. The Q.A. Manager could be very helpful explaining some of the technical details either in the meeting or outside it.

By profiling the audience, you can anticipate some of the issues that are likely to come up. This helps you with figuring out a script for dealing with the issues. The issues are likely to surface during the question and answer session. Audiences are becoming less patient; another trend is they want more interaction during the presentation. The

Table 2.3. Example of Using Presentation Preparation Form (continued)

Likely Questions	Answers
1. The cost of training ($50,000) seems high. How can you justify this?	(a) Quantify the cost/benefits of using DOE for product introductions in near future. Benefits: Sooner to market Less development costs Cost: DOE training (b) Calculate payback period and rate of return.
2. How can you get more information with less samples?	(a) Show brief explanation of matrix symmetry using viewgraph of cube for 3-dimensional matrix. (b) Details will be covered in training. May need help from QA Manager to explain certain concepts. (c) Drawback is extra preparation necessary for planning samples. But with this simple preparation, can get benefits listed in Main Points.
3. Why are you bringing up this idea rather than the QA Manager or Engineering Director?	(a) We all share the interest of making this organization more successful. (b) If can achieve benefits listed in Main Points, it shouldn't matter who proposes the idea. (c) Key point: work as a team to get this proposal off the ground and enjoy its benefits.

question and answer session is the first opportunity the audience gets to interact.

Anticipate the most difficult questions and controversial areas that may come up. Then write down how you will answer these questions. The key here is to answer negative questions honestly and concisely and then link back to your main points—your selling points. For instance, answer the difficult question, then use a "but" or "however" to get focus back to your main points. In the example listed in Table 2.3, the answer to the question about justifying the high cost of training may go something like this.

"You are right the costs are high, but they need to be compared to the benefits mentioned in my talk. The benefits are getting to market sooner and lower product introduction costs. I estimate the rate of return for the training costs to be..."

As you work through your answer, gather the necessary background information and come to the presentation prepared with your answers. Answers to complicated questions may include separate viewgraphs.

Answer questions in a succinct manner. Be candid about not knowing an answer, but if you don't know an answer, commit to getting the answer.

Using Table 2.2 to prepare your presentation forces you to think about the audience and where the controversial areas are. Many times it is beneficial to discuss your presentation and ideas with key personnel individually before your presentation. In this way you can gain better understanding of the problem areas and you can also gain support for your ideas. For the example shown in Table 2.3, talk to the Engineering Director and Quality Assurance Manager before your presentation. By talking with the Engineering Director, you may be able to address some detailed, technical questions beforehand. Also you can work on the idea together, so the Engineering Director will be less defensive and more supportive. By talking with the Q.A. Manager, you can gain support for your idea and prepare her in case she needs to answer some of the technical questions.

The form in Table 2.2 serves as the backbone for your presentation preparation. At this point you have identified the purpose and main points of your talk, profiled the audience, and anticipated how to handle the most difficult questions. This preparation gives you clues as to how best develop your ideas during the middle part of the talk. The most compelling ideas are built on clear, concrete facts. Keep your explanations short. Present your points visually whenever possible by using props, charts, or drawings. Keep your visual aids simple. The text used for any visual aids should be large enough to be read easily from the back of the room. Use a font size of at least 20 points if possible. The text used in overhead transparencies should not be a dissertation. Write succinctly main ideas and points on the overheads. Then embellish these visual aids with your verbal explanations during the presentation.

The last part of your preparation is to rehearse your presentation out loud. You can do this in the room where the presentation will take place, so you can get a better feeling of the physical setting and space of the room. Or you can practice in front of a mirror or in front of colleagues who can offer suggestions for improvement.

As mentioned earlier, being able to make effective presentations is an essential interpersonal communication skill and is important for career advancement. Being an effective presenter is a learned skill. Remember to keep your presentations focused and take the time to prepare in order to make the biggest, positive impression upon your audience.

Chapter 3

PRACTICAL IMPLEMENTATION

So far you have been shown pointers on how to improve personal and interpersonal communication. Now it is time to use these ideas to get some practical, immediate results.

1. HOW TO GET ALONG WITH YOUR BOSS

Today's organizations are becoming more complex. You need to take on more responsibility for your success and for the organization's success. The best way to start taking the initiative is to build a good working relationship with your boss. A good working relationship with your manager is fundamental to your success. It helps you work more effectively in the short-term and is a wise investment of your time in the long-term. You need your manager's support in order to be effective. The importance of your manager demands your attention. Take the initiative to make your relationship work.

The foundation for building a good relationship with your manager (see Table 3.1) ties in with the principles for good interpersonal communication (Table 2.1). First and foremost, you and your manager must respect and trust each other. This does not mean you have to love or even like your manager, but you do need respect and trust. Ways to build trust were covered in the section describing Table 2.1. Be honest. Keep your promises faithfully. Make sure your actions support your communications. In this way, you communicate for the long-term as well as for today.

**Table 3.1. Foundation for a Good Relationship
With Your Manager**

---

1. Respect and trust

2. Mutual understanding of what is important (goals and priorities)

3. Good communication about day-to-day activities

4. Share common interest to improve organization's success

The second item for building a good relationship with your manager is to make sure you have a mutual understanding of the goals and priorities for yourself and the organization. Set up weekly, one-on-one meetings with your boss to review your priorities, goals, and progress made. This ensures you are staying on track to do the right things and ensures you have your manager's complete attention for a certain time each week. Make sure you ask questions if you don't know or understand something. Also find out about your manager's priorities. Identify ways to help your manager with their priorities. One way of helping is gathering information your manager needs to make better decisions. Another way is to pitch in to do work your manager either does not like to do or does not do well. Above all, end each meeting with your manager by identifying a follow-up plan. Agree on who will do what by when, and establish the time for a follow-up meeting.

The next item for building a good relationship with your manager is to have good communications regarding day-to-day activities. The trick is to keep your manager informed without wasting his or her time. Your manager's time is a limited resource. Use it wisely and get to know your manager's style and preferences for staying informed. An important point is to communicate to your manager problems encountered with your work before he or she hears about them from others. Managers do not like negative surprises about their own group coming from outside the group. Avoid going around your manager and communicating with your manager's boss. If you do communicate with your manager's boss, keep your manager informed. Also ask your manager for feedback on how specific things are going. This helps ensure you are doing the right things right. Utilize the principles described earlier about how to listen and setting up feedback systems. In order not to overwhelm your manager, ask for feedback on one specific area of performance at a time. Remember your manager is one of your most important customers.

The last point in building a good relationship with your manager is to share a common interest to improve the organization's success. This is your common bond. Even though you may disagree from time to time on certain points, you can always agree you have the group's best interests in mind. Avoid being a pessimist. When problems arise, try to provide options to solve the situation instead of just presenting the problem. Focus on improving the group's performance, and don't dwell on personality differences.

Establishing a good working relationship with your boss is invaluable to your professional success. You not only gain important insights into where your company is heading, but you also pave the way to discuss the professional side of your personal strategic plan. Without knowing where you want to go, your boss cannot help you get there. Talk about your professional goals and development plan with your manager. Your manager can help you get assignments and training to meet your goals.

2. HOW TO GET ALONG WITH YOUR CO-WORKERS

This is another application of your communication skills that is vital to your success. It requires utilizing what you have learned to improve your interpersonal communication, how to listen, setting up feedback systems, and presenting your ideas.

Take initiative to build good relationships with your co-workers. As organizations are becoming more interdependent, cooperation and collaboration among co-workers is essential. Be a team player. Look out for the needs of others as well as your own. Taking the time to build good relationships with your co-workers not only makes your work environment more enjoyable, it can help you and your organization get results.

Steps to build good relationships with your co-workers are listed in Table 3.2. The first step is to identify the co-workers who are most critical for your workplace success. These are typically your suppliers and customers, both internal and external. Define the information, product, or service that is exchanged between you and each co-worker,

---

**Table 3.2. Building Good Relationships
With Your Co-Workers**

1. Identify "significant" co-workers.
2. Take initiative to talk with your co-workers.
3. Explore common points of interest.
4. Identify ways to help each other.
5. Agree on a specific follow-up plan.

---

and see how this fits in with your goals and purpose. This exercise will help you define your most important co-worker relationships. Start with the two or three most important relationships. Evaluate the nature of the relationship at present. This can range anywhere from a trusted friend to a person difficult to work with.

Next take the initiative to improve the relationship. Set up a time where the two of you can meet, face-to-face, without interruptions. One convenient way is to invite your co-worker out to lunch. This is especially appropriate after the co-worker has done something which really helped you (such as giving you good advice, reviewing your project plan for you, or finishing their work ahead of schedule). This is a low-key method to converse in a more relaxing atmosphere than at someone's desk. If lunch is not possible, try setting up a meeting where you will not be interrupted by phone calls or other people. If the co-worker seems reluctant or suspicious, be persistent. Make it clear how important you view the relationship for both you and the co-worker. Stress the main purpose of meeting is so you can better understand the needs of your co-worker and identify how you can better help each other.

When you meet, start by exploring common points of interest. Since most likely you have some sort of customer-supplier relationship or work together as a supplier to a common customer, one common point could be the information, product, or service exchanged. Identify the common organization you belong to, if applicable. Understand each other better by discussing the goals and priorities for both of you. Understand how the common information, product, or service is important in relation to these goals and priorities. This shows your importance to each other and how beneficial your collaboration can be.

Then identify ways you can help each other. Items for discussion could be:

- how you can do your jobs better
- how the organization can work better
- where you think the organization is going from a strategic standpoint

The last step is to agree on a specific follow-up plan. It is important to start with a simple plan which is easy to implement. As you work to improve a co-worker relationship, you want to build a sense of trust with one another. By keeping the plan easily doable at the start, you can make sure to keep your promises faithfully. More difficult tasks can be entertained as the relationship grows. As part of the specific follow-up plan, identify who will do what by when, and set up a time for a follow-on discussion. Finally express your appreciation with your co-worker for taking the time and effort to build or maintain a good relationship.

Such meetings if handled correctly can win you powerful allies. It will strengthen your relationships with co-workers and give you valuable insights into your organization. You will get many good ideas. Make sure you offer ideas which will help your co-workers meet some of their goals. By building such relationships, co-workers will go out of their way to help you when asked. It may also win you good friends and make your job more fun.

3. CONFRONTING PROBLEMS WITH YOUR BOSS OR CO-WORKERS

The last example is probably the most difficult application of interpersonal communication. This is the sternest test for using the principles of good communication, how to listen, and getting your ideas across. To the extent you have already taken the initiative to establish a good working relationship with your boss and co-workers, the easier it is to confront them with a problem. The more respect and trust you have for each other, the smoother and more successful your encounter will be.

Consult with your boss or co-worker when you have a problem. The problem may be a conflict of goals and priorities. It may be a situation or decision made that is hurting your work performance. Or it may simply be you are upset with your boss or co-worker.

There are three ways to handle a problem with your boss or co-worker.

1) Ignore the problem.

2) Try to solve the problem indirectly.

3) Try to solve the problem directly.

The first two ways will not solve the problem. The best way to solve the problem is to work directly with your boss or co-worker to solve the problem in a mutually agreeable manner. It is also desirable to handle problems early on before they develop into major conflicts.

Thus the first step in dealing with a problem with your boss or co-worker is to take initiative to solve the problem (see Table 3.3). Set up a time to discuss the problem in private. It is not wise to bring up problems in public. If brought up in public, there is a greater risk the other person may get embarrassed, lose self-esteem, and become defensive. Discuss the problem in private to minimize the emotions involved.

Begin your discussion by describing the problem from your point of view. Focus on how the problem is impacting your performance. Be specific. If appropriate, mention how the situation makes you feel. Try not to focus on the personalities involved. Keep the description of the problem brief. Ideally the description should take one minute or less. Be brief because as you are describing the problem, the natural tendency is for the emotions of the recipient to increase dramatically. Give the person a chance to respond before their pent-up emotions become detrimental to the discussion.

After briefly describing the problem, ask the other person how he or she sees the problem from their point of view. Utilize all your listening skills to truly understand the other person's frame of reference. Always try to understand both sides of the situation before solving a problem.

Once you understand each other in terms of the problem, begin to work collaboratively to solve the problem. First agree on what is the specific problem. You may have to restate the problem a number of different ways before you get agreement. Next identify ways to solve

Table 3.3. Confronting Problems With Your Boss or Co-Workers
1. Take initiative to solve the problem.
2. Meet with the other person alone.
3. Describe the problem from your point of view.
a) Be brief.
b) Focus on how it impacts your performance; don't focus on personalities.
4. Listen to understand the other person's point of view.
5. Agree on the problem to solve.
6. Identify ways to solve the problem.
7. Agree on a specific follow-up plan.

the problem. Avoid evaluating the proposed solutions at first. The goal is to get as many ideas out on the table. Once all the ideas you can collectively think of have been proposed, then start evaluating them. Determine the best solution.

And finally agree on a specific follow-up plan. This is an important step for any meeting where you are trying to build relationships or solve problems. The follow-up plan gives both of you a sense of accomplishment and direction coming from the discussion. Determine who will do what by when, and arrange a time for a follow-up meeting to check progress.

KEY POINTS: Communications

1) Communicate with yourself. Develop a personal strategic plan. This plan should include your lifetime goals, a development plan to achieve these goals, and an analysis of your environment.

2) Manage your time on a daily basis by focusing on your most important tasks. Spend a portion of your time every day working toward accomplishing goals in your personal strategic plan.

3) Good interpersonal communications are built on understanding the viewpoint of others. Utilize listening skills to understand the other person's frame of reference. Set up feedback systems with your internal and external customers.

4) Take the time to plan and prepare for your face-to-face communications. Define the purpose and main points. Profile your audience. Anticipate questions or concerns that may come up. End the discussion by summarizing the key points and define a specific follow-up plan.

5) Good interpersonal relationships are built upon trust. Take the initiative to build good working relationships with your boss and your co-workers. Handle problems with them by speaking directly to them and working together to develop a solution.

SUGGESTED ACTIVITIES: Communications

1) Formulate a personal strategic plan by following the steps listed in Table 1.1.

   a) Write down your lifetime goals. Consider goals both inside and outside of work. Share them with others close to you. Rewrite them within 72 hours.

   b) Analyze your capabilities in relation to skills needed to achieve your lifetime goals. Identify capabilities you need to develop, and devise a plan how you will develop them.

   c) Break down your lifetime goals into shorter term goals.

   d) Analyze your environment to determine if it is conducive to achieving your development plan and lifetime goals.

Consider modifying your environment or your goals if there is a mismatch.

2) Analyze how you use your time for a period of two weeks. Compare your actual time usage to the time usage necessary to achieve your lifetime goals. Make adjustments if needed. Set aside some time everyday dedicated toward achieving your lifetime goals.

3) Identify who your customers are at work. List those people inside and outside your organization who depend on you for various goods, services, projects, or information. Set up feedback systems with them to find out their current level of satisfaction and how their satisfaction can be improved. Take steps to increase the satisfaction of your customers.

4) Use Table 2.2 to prepare for your next presentation. Write down the purpose and main points, profile your audience, and write down answers to likely questions.

5) Establish or maintain a good working relationship with your boss. Key items are listed in Table 3.1. Set up weekly one-on-one meetings with your boss to review priorities, goals, and progress made. Emphasize your common interest to improve the organization's success.

6) Establish or maintain good working relationships with your co-workers (see Table 3.2). Take the initiative to talk with them, explore common points of interest, and identify ways to help each other.

7) Use the steps listed in Table 3.3 to address a problem you are currently facing with your boss or a co-worker. Work directly with the other person to solve the problem. Remember to focus on how the problem impacts your performance, and don't focus on personalities.

REFERENCES: Communications

Goal Setting/Time Management

Bennis, W., *On Becoming a Leader,* Reading, Massachusetts: Addison-Wesley Publishing Company, 1989.

Blanchard, K. H., *One Minute Manager Meets the Monkey*, New York: William Morrow & Company, 1989.

Covey, S. R., *Principle-Centered Leadership*, New York: Simon & Schuster, 1990.

Covey, S. R., *The 7 Habits of Highly Effective People*, New York: Simon & Schuster, 1989.

Douglass, D. N. and Douglass, M. E., *Manage Your Time, Your Work, Yourself*, New York: A M A C O M, 1993.

Ferner, J. D., *Successful Time Management*, New York: John Wiley & Sons, 1995.

Greeson, G., *Goal Setting: Turning Your Mountains Into Molehills*, Saint Charles: Potential Unlimited, 1994.

Helmer, R. G., *Time Management for Engineers & Constructors*, New York: American Society of Civil Engineers, 1991.

Jones, J. J., *Goal Realization: The Project Management Objective*, New Haven: P D S Publications, Incorporated, 1993.

Lakein, A., *How to Get Control of Your Time & Your Life*, New York: NAL/Dutton, 1989.

Merrill, A. R., Covey, S. R., and Merrill, R. R., *First Things First: A Principle-Centered Approach to Time & Life Management*, New York: Simon & Schuster Trade, 1994.

Moskowitz, R., *How to Organize Your Work and Your Life, 2nd Edition*, New York: Doubleday & Company, 1981.

Roquemore, E., *How to Develop a Goal Mind: Using Left & Right Brain Functions to Achieve Your Goals*, Whittier: Roquemore Seminars, 1995.

Skopec, E. W. and Kiely, L., *Taking Charge: Time Management for Personal & Professional Productivity*, Reading: Addison-Wesley Publishing Company, Incorporated, 1991.

Smith, B. and Sher, B., *I Could Do Anything If I Only Knew What It Was: How to Discover What You Really Want & How to Get It*, New York: Delacorte Press, 1994.

Wilson, S. B., *Goal Setting*, New York: A M A C O M, 1994.

Oral Communication

Axtell, R. E., *The Do's & Taboos of Public Speaking: How to Get Those Butterflies Flying in Formation*, New York: John Wiley & Sons, Incorporated, 1992.

Berg, K. and Gilman, A., *Get to the Point: How to Say What You Mean and Get What You Want*, New York: Bantam Audio Publishing, 1989.

Booher, D., *Communicate with Confidence!: How to Say It Right the First Time & Every Time*, New York: McGraw-Hill Companies, 1994.

Corporate Classrooms Staff, *Prentice Hall's Get a Grip on Speaking & Listening: Vital Communication Skills for Today's Business World*, Englewood Cliffs: Prentice Hall, 1995.

Elgin, S. H., *BusinessSpeak: Using the Gentle Art of Verbal Persuasion to Get What You Want at Work*, New York: McGraw-Hill Companies, 1995.

Gatto, R. P., *A Practical Guide to Effective Presentation*, Pittsburgh: G T A Press, 1990.

Vasile, A. J. and Mintz, H. K., *Speak with Confidence: A Practical Guide, 7th Edition*, New York: HarperCollins College,. 1995.

Whetten, D. A. and Cameron, K. S., *Developing Management Skills,* Glenview, Illinois: Scott, Foresman and Company, 1984.

Listening

Baker, H., Jr., *One Minute Listener*, Salem: Forum Press International, 1991.

Banville, T. G., *How to Listen-How to Be Heard*, Chicago: Nelson-Hall, Incorporated, 1978.

Dugger, J., *Listen Up: Hear What's Really Being Said*, Overland Park: National Press Publications, 1992.

Helgesen, M. and Brown, S., *Active Listening: Building Skills for Understanding (Student's Book)*, New York: Cambridge University Press, 1993.

Lougheed, L., *Listening Between the Lines: A Cultural Approach*, Reading: Addison-Wesley Publishing Company, Incorporated, 1987.

Roberton, A., *Listen for Success: A Guide to Effective Listening*, Burr Ridge: Irwin Professional Publishing, 1993.

Whetten, D. A. and Cameron, K. S., *Developing Management Skills,* Glenview, Illinois: Scott, Foresman and Company, 1984.

Communication of Technical Information

Adamy, D. L., *Preparing & Delivering Effective Technical Presentations*, Norwood: Artech House, Incorporated, 1987.

American National Standard for Guidelines for Organization, Preparation & Production of Scientific & Technical Reports, New York: American National Standards Institute, 1982.

Barnum, C. M., *Prose & Cons: The Do's & Don'ts of Technical & Business Writing*, Englewood Cliffs: Prentice Hall, 1986.

Beer, D. F., Editor, *Writing & Speaking in the Technology Professions: A Practical Guide*, Piscataway: Institute of Electrical & Electronics Engineers, Incorporated, 1992.

Bell, A. H. and Damerst, W. A., *Clear Technical Communication: A Process Approach, 3rd Edition*, Orlando: Dryden Press, 1989.

Bolsky, M. I., AT&T Bell Laboratories Staff, *Better Scientific & Technical Writing*, Englewood Cliffs: Prentice Hall, 1988.

Brogan, J. A., *Clear Technical Writing*, New York: McGraw-Hill Companies, 1982.

Brown, J. F, Laws, L., Editor, and Bowers, C., Editor, *A Student Guide to Engineering Report Writing, 3rd Revision*, Solana Beach: United Western Press, 1989.

Burnett, R. E., *Technical Communication, 3rd Edition*, Belmont: Wadsworth Publishing Company, 1994.

Cain, B. E., *The Basics of Technical Communicating*, Washington: American Chemical Society, 1988.

Casagrande, D. O. and Casagrande, R. D., *Oral Communication in Technical Professions & Businesses*, Belmont: Wadsworth Publishing Company, 1986.

Dagher, J. P., *Technical Communication: A Practical Guide*, Englewood Cliffs: Prentice Hall, 1978.

ELS, Inc. Staff, *Career English Engineering: Civil & Mechanical*, Boston: Heinle & Heinle Publishers, Incorporated, 1984.

ELS, Inc. Staff, *Career English Engineering: Electrical*, Boston: Heinle & Heinle Publishers, Incorporated, 1984.

Gray, J. G., Jr., *Strategies & Skills of Technical Presentations: A Guide for Professionals in Business & Industry*, Westport: Greenwood Publishing Group, Incorporated, 1986.

Haines, R. W. and Bahnfleth, D. R., *Effective Communications for Engineers*, Blue Ridge Summit: T A B Books, 1989.

Kirkman, J., *Good Style: Writing for Science & Technology*, New York: Chapman & Hall, 1992.

Mablekos, C. M., *Engineer's Guide to Business-Presentations That Work*, Piscataway: Institute of Electrical & Electronics Engineers, Incorporated, 1991.

Mablekos, C. M., Editor and Plotkin, H. E., *Technical Writing & Communication*, Washington: American Chemical Society, 1988.

Mandel, S. and Gerould, W. P., Editor, *Technical Presentation Skills*, Menlo Park: Crisp Publications, Incorporated, 1994.

Marlow, A. J., *Technical Documentation, 2nd Edition*, Cambridge: Blackwell Publishers, 1995.

McKown, D., *Engineer's Guide to Business-Writing for Career Growth*, Piscataway: Institute of Electrical & Electronics Engineers, Incorporated, 1991.

Michaelson, H. B., *How to Write & Publish Engineering Papers & Reports, 2nd Edition*, Philadelphia: I S I Press, 1986.

Miles, T. H., *Critical Thinking & Writing for Science & Technology*, Orlando: Harcourt Brace College Publishers, 1990.

Morrisey, G. L. and Sechrest, T. L., *Effective Business & Technical Presentations, 3rd Edition*, Reading: Addison-Wesley Publishing Company, Incorporated, 1987.

Pattow, D. and Wresch, W., *Communicating Technical Information: A Guide for the Electronic Age*, Englewood Cliffs: Prentice Hall, 1992

Robinson, P. A., *Fundamentals of Technical Writing*, Boston: Houghton Mifflin Company, 1985.

Roundy, N. L. and Mair, D., *Strategies for Technical Communication*, New York: HarperCollins College, 1987.

Smith, L. R., *Basic English for Business & Technical Careers*, Englewood Cliffs: Prentice Hall, 1985.

Whalen, T., *Writing & Managing Winning Technical Proposals, 2nd Revision*, Vienna: Holbrook & Kellogg, 1994.

White, E. B. and Strunk, W., Jr., *Elements of Style, 3rd Edition*, 1989.

Woelfle, R. M., Editor, *A New Guide for Better Technical Presentations: Applying Proven Techniques with Modern Tools*, Piscataway: Institute of Electrical & Electronics Engineers, Incorporated, 1992.

Woolston, D. C., Robinson, P. A., and Kutzbach, G., *Effective Writing Strategies for Engineers & Scientists*, Lewis Publishers, 1988.

Working with Your Boss

Barner, R., *Lifeboat Strategies: How to Keep Your Career above Water During Tough Times - or Any Time*, New York: A M A C O M, 1993.

Bing, S., Rubenstein, J., Editor, *Crazy Bosses*, New York: Pocket Books, 1993.

Bramson, R. M., *Coping with Difficult Bosses: Dealing with Ogres, Wafflers, Know-it-alls, & Paranoids that Run the Office*, New York: Simon & Schuster Trade, 1993.

Des Roches, B., *Your Boss Is Not Your Mother: Breaking Free from Emotional Politics to Achieve Independence & Success at Work*, New York: Avon Books, 1995.

Hegarty, C., *How to Manage Your Boss*, New York: Ballantine Books, Incorporated, 1985.

Kennard, K., *How to Manage Your Boss: Build a Partnership with Your Boss That Will Make You Both Successful*, Overland Park: National Press Publications, 1991.

Lovett, A. B., *Career Prescription: How to Stop Sabotaging Your Career & Put It on a Winning Track*, Englewood Cliffs: Prentice Hall, 1994.

Shainis, M. J., "How to Get Along with Your Boss," *Consulting Engineer,*
    September 1982.

Conflict Management

Arnold, J. D., *When the Sparks Fly: Resolving Conflicts in Your Organization,*
    New York: McGraw-Hill Companies, 1993.

Hendricks, W., *How to Manage Conflict: How to Handle Difficult People &
    Situations for Win-Win Results,* Overland Park: National Press Publications,
    1992.

Murphy, J., *Managing Conflict at Work,* Burr Ridge: Irwin Professional
    Publishing, 1993.

Weeks, D., *The Eight Essential Steps to Conflict Resolution,* New York: Putnam
    Publishing Group, 1994.

Wisinski, J., *Resolving Conflicts on the Job (A Worksmart Book),* New York: A M
    A C O M, 1993.

PART II

STRATEGIC PLANNING

Defining the Path to Success for Your Organization

Strategic planning defines the backbone of the organization. The strategic plan for an organization is similar to the personal strategic plans discussed in Part I. They both consider three elements: goals, capabilities, and environment. The strategic plan for an organization examines the goals of the organization, the capabilities of the organization, and the business environment in order to outline the steps for the organization to achieve long-term success. It balances a long-term vision with short-term considerations. The result of having a well-defined and well-understood strategic plan is knowing where your company is going and understanding your role in accomplishing this plan.

The purpose of Part II is to define the components of effective strategic planning and to show how engineers can assist with the strategic planning process. The executives have the responsibility to develop the strategic plan. Your role as an engineer is to supply them with information to help them make better decisions when formulating the plan. Another responsibility you have is to thoroughly understand what is in your organization's strategic plan. This is important whether you are starting a new job or have been with the organization for many years. Consequently you can see how all your activities funnel into the overall plan for the organization. Judge constantly what you should be working on versus what is convenient to work on. Understanding the strategic plan and working on the right things maximizes your worth to the organization.

There are three major problems with not having a well-defined strategic plan. These problems occur all too often in the engineering world today.

1) **The business focus is short-term.**
   This means the main priorities are to meet the sales numbers or budget expenses for the month. Your work priorities change on a quarterly basis before you can accomplish the "number one" priority you had in past months. Engineers are unsure of what is important, and the result is a loss of morale.

2) **There is a lack of coordination among groups within the organization to achieve its overall goals.**
   The attainment of the most important goals for an organization invariably requires teamwork among the various functional groups to be successful. Without a well-defined plan, many times executives do not agree on the

organization's priorities. The result is that each group tries to accomplish what they believe individually to be the most important goals. This works as long as each group can operate autonomously without resources from each other. In practice certain resources always need to be shared among groups. Without a common vision, teamwork suffers.

3) **There is a failure to recognize the tradeoffs resulting from resource limitations.**
   The strategic plan needs to define the most important tasks for success of the organization. Only these tasks should be worked on. The most important resources in any organization are its people. There are always limitations to this resource. Thus, executives need to define the few tasks which are vital for success and keep the people in the organization focused on these tasks. They also need to recognize that other less important tasks will not receive attention. Success comes only through concentration of effort and effective execution. If the people within the organization are expected to perform as well on the less important tasks as they do on the most important tasks, the end result is performance which is universally mediocre.

One of the most important responsibilities of executives is to define a strategic plan for the organization. A well-defined plan means there is proper balance and attention paid to both long-term and short-term issues. There is consistency in what the organization is striving for, and the major tasks for accomplishing this plan follow a logical progression. These major tasks are understood by everyone in the organization, and this understanding helps develop cooperation among working groups towards accomplishing common goals.

In practice, strategic planning is a dynamic process that requires continual updating as the business environment changes. As an engineer, part of the frustration in understanding the strategic plan is that it is constantly changing. However two necessary components of a strategic plan are: stating the assumptions and developing contingency plans. Thus even possible changes should be planned in advance as much as possible so that the organization is acting proactively rather than reactively.

Whether or not your organization has a crisp, well-understood strategic plan, it is your obligation to exhibit to the executives that you recognize the importance of strategic planning. Executives set the strategic plan and vision for the organization, but you can help formulate this plan. As an engineer working in the trenches, you have a perspective the executives often do not appreciate when defining the plan. This perspective affords you with unique insight to give inputs into what new products could be developed or where costs could be reduced. You can also provide feedback regarding resources and costs required to achieve certain tasks in the strategic plan, and you may have other ideas how the goals can be achieved more efficiently and faster. Take on the role to help write the strategic plan. After all, having a strong strategic plan is essential not only for the organization's success but also for your success.

Chapter 4

DEFINITION OF STRATEGIC PLANNING

In an increasingly uncertain world, the business environment is becoming more complex and subject to rapid change. Firms worldwide are facing more competition in both domestic and international arenas. Strategic planning is the process of relating an organization to its environment. This process has great value. It defines a path for the organization to be successful. However there is a tendency to treat the strategic plan as strictly a short-term, quantitative business plan. To be most effective, the strategic plan needs to balance short-term and long-term issues and analyze quantitative and qualitative aspects.

The strategic plan examines three areas: the goals of the organization, the capabilities of the organization, and the business environment. This is shown in Figure 4.1. The strategy is to define a course of action taking into account each of these areas to create and maintain a competitive advantage. This competitive advantage leads to creating and maintaining above average profitability.

1. GOALS

Goals define what an organization would like to do. Goals should clarify organizational targets that are qualitative, such as defining the organization's values, the products or services it provides, and its target market, and quantitative such as financial goals.

Figure 4.1. Elements of a Strategic Plan

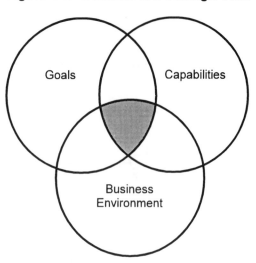

1.1. Values

Values define what your organization stands for. Even though other areas of the strategic plan change such as the business environment, capabilities, and other aspects of the organization's goals, the organization's values should seldom change if ever. They represent the core ideals of the organization, and they give rise to the company's culture. You understand the reference from which you need to operate by knowing your organization's values.

One value is defining the organization's identity. As an organization with engineers, does it want to be a technical leader or a technical follower? How important is research and development? How important is manufacturing capability? Is the organization aggressively pursuing growth or is it more conservative? How important are the people for the organization to achieve its goals? Setting the organization's identity requires outlining its values.

Another value is the organization's attitude toward risk. How much risk is the organization willing to tolerate in order to achieve its financial goals? In general, more risk must be tolerated as the desired sustainable rate of growth increases. In order to achieve high rates of growth, it is often necessary to increase investment into the business. This involves higher risk. Organizations must be honest with the amount of risk they are willing to take. The financial goals and the attitude toward risk must match in order for the organization to be successful. Otherwise the strategic plan is not realistic.

Defining the organization's identity is beyond the control of engineers. It is the responsibility of executives to determine these values. But you need to understand the organization's values, and because values are qualitative, they are sometimes difficult to determine. To understand your organization's values, the first place to start is to read your organization's strategic plan. Ask your manager for the organization's strategic plan, and as you read it, look for statements pertaining to values.

The next step is to "read between the lines" by talking with others. An excellent way to do this is by asking your manager about the organization's values during your one-on-one meetings. Also ask the opinions of your co-workers as you establish working relationships (discussed in Part I). It helps to get a number of perspectives to gain a better understanding. This understanding of the organization's values can give you guidelines how you can contribute to your organization's success.

1.2. Products and Services

Strategic planning defines the types of products and services the organization wants to provide. These are the tangible outputs after considering the organization's values. These products and services must possess advantages over the competition in order to help the organization achieve its financial goals.

A technical leader must be able to produce innovative products. These products must satisfy customers in ways other products have not. A technical follower must be able to produce products that also satisfy a customer need. The product may be a "me-too" product, but it must have a competitive advantage. This could be in terms of performance, price, service, or availability.

1.3. Target Market

The goals also include a definition of the target market. The target market is a description of the desired customers for your products or services. There must be a good understanding of how the product satisfies a customer need in order for the organization to be successful. This includes defining the goals for your organization's position within the target market versus the competition. There should be a goal for market share. Market share is the percentage of your organization's sales in the target market compared with total target market sales.

1.4. Financial Goals

One most often thinks of goals in terms of financial goals because these are easily communicated. The financial goals include targets for sales, profits, and sustainable rate of growth. Sales are the amount of revenue the business receives from selling goods and services to customers. Profits are a measure of the revenue remaining after all expenses have been subtracted. Sustainable rate of growth is an indicator of how much the money invested in the business will grow annually. These goals in the strategic plan become key financial measures for the organization. It is important for you to understand what the key financial measures are and how you can impact them. An explanation of key financial measures and how you can link your activities to have a positive effect on them will be discussed in Part III.

2. CAPABILITIES

A comparison must be made between the organization and other firms in the industry in order to assess the organization's strengths and weaknesses. These strengths and weaknesses can be broken down by functional areas of the organization such as Engineering, Manufacturing, Marketing, and Customer Service.

The capabilities of the organization must match the skills necessary to achieve the organization's goals. For instance if the organization wants to be a technical leader, it must have a strong research and development effort. Capabilities to develop innovative products and getting them to the market quickly are required. If the organization is to compete as a cost leader, it must have very efficient manufacturing capability.

In the case where the organization's goals require capabilities where it is weak, there must be a sound plan to develop or purchase the necessary skills. Otherwise the strategic plan is not realistic. The organization's goals and capabilities must match.

3. BUSINESS ENVIRONMENT

To analyze the business environment, the organization must assess the external forces of its industry. As mentioned earlier, the purpose of strategic planning is to define a course of action to match the organization's goals and capabilities to its environment. The problem is that the business environment is continually changing. In strategic planning, the environment warrants the most attention because it cannot usually be controlled as easily as goals and capabilities. Therefore an uncertain world creates the need for strategic planning, and a main part of the planning process is examining the environment.

There are increasing risks associated with operating a firm in an increasingly uncertain world. The purpose of strategic planning is not to avoid risk but to manage risk. In the long run, the firms earning above average profits are the ones that take intelligent risks to create and maintain a competitive advantage. To take intelligent risks, the structure of the environment must be analyzed by determining the strength of five competitive forces as described by Porter. These are shown in Table 4.1. The following describes these competitive forces and the impact they have on the profitability of firms competing in the industry.

The intent of describing these competitive forces is to help you understand strategic planning. Executives have the responsibility to

Table 4.1. Competitive Forces Determining Industry Profitability[5,6]
1. Threat of new entrants
2. Threat of substitute products or services
3. Bargaining power of buyers
4. Bargaining power of suppliers
5. Intensity of rivalry among existing firms

5. Michael E. Porter, *Competitive Strategy: Techniques for Analyzing Industries and Competitors,* New York: The Free Press, 1980.

6. Michael E. Porter, *Competitive Advantage: Creating and Sustaining Superior Performance,* New York: The Free Press, 1985.

assess the business environment, but by knowing about competitive forces, you have a better understanding of the framework used to assess the business environment. You can also analyze the competitive forces of your industry, from your standpoint, to help assess the future of the industry in which you are working.

3.1. Threat of New Entrants

The threat of new entrants is the threat of new firms entering the industry. This affects profits because new entrants mean increased competition. There can be two effects as a result of new entrants. The selling price of your product or service decreases, or the cost to produce your product or service increases. Both these effects lower profits. Barriers to entry such as economies of scale, proprietary technical knowledge, or expected retaliation from existing firms are needed to reduce the threat of new entrants.

3.2. Threat of Substitution

Similarly, pressure from substitute products tends to drive down profits toward free market levels. Substitute products are those that perform the same function as your product. As an example, substitute products confronted the manufacturers of mainframe computers. As the technology advanced, workstations and personal computers were able to replace many of the functions that previously required a mainframe computer.

The most threatening substitute products improve the performance-price ratio for the customer compared to the existing product. Another threat of substitution occurs when a highly profitable firm enters the market with a substitute product. This is threatening because a highly profitable firm is in a stronger position to take more risks and provide more resources to enhance the sales of its substitute product.

3.3. Bargaining Power of Buyers

Buyers are consumers of the product or service sold by individual firms in the industry. The bargaining power of buyers affects an industry's profits. Buyers pit individual firms of an industry against each other when they ask for lower prices and higher quality. Powerful buyers can squeeze the profitability out of firms vying for their business.

Powerful buyers purchase large volumes of product relative to the sales volume of the individual firms in the industry. When this is the case, the fate of the industry rests in the hands of the buyers. These buyers are then in a strong position to negotiate lower prices. Also powerful buyers purchase products which are largely undifferentiated. The products are commodity products. Here the buyers can compare prices from each of the competing firms. The buyers are then in a strong negotiating position.

An interesting example of the power of buyers and how the business environment can change is the purchasers of DRAM memory chips used in personal computers. These memory chips are largely undifferentiated among the various manufacturers. These chips are commodity products, and there have been tremendous advances in integrated circuit processing techniques. This led to precipitous declines in their prices. The average selling price for a 1-Megabit DRAM declined from $24.88 in 1986 to $2.99 in 1992. During this time, the buyers of DRAMs had power over DRAM manufacturers, and it was difficult for the manufacturers to be profitable. However in more recent years, the market for DRAMs changed. As engineers have developed denser and more powerful chips, DRAM use spread not only in computers, but to other products such as cellular phones, pagers, camcorders, and microwave ovens.[7] The DRAM market grew by an average rate of 40% during 1992 to 1995, and the DRAM demand exceeded the supply. With the market diversifying, individual DRAM buyers no longer purchased large volumes of product relative to the sales volume of the individual manufacturers. This meant the DRAM buyers no longer had purchasing power over manufacturers. The result was DRAM prices actually increased from 1992 to 1995, and the DRAM manufacturers were very profitable during this period.

Then the business environment changed dramatically once again. The high profits made by the DRAM manufacturers led to more investment into wafer fabrication plants in order to increase manufacturing capacity for DRAMs. DRAM supply increased

7.   Julie Chao, and David P. Hamilton, "Bad Times Are Just a Memory for DRAM Chip Makers," *The Wall Street Journal,* Aug. 28, 1995.

significantly and exceeded demand. Prices plummeted. The price for 16-Megabit DRAMs fell more than 80% in 1996, and the price for 64-Megabit devices declined from $250 at the beginning of 1996 to $38.50 by late April, 1997. DRAM manufacturers lost much money during this most recent period, and many of them are now de-emphasizing the importance of DRAMs in their product lines. DRAMs are an example of how fast the business environment can change.

3.4. Bargaining Power of Suppliers

Suppliers sell components or services to individual firms in the industry in order for the individual firms to produce their product. Suppliers to the industry exert bargaining power by threatening to increase prices or reduce the quality of purchased goods. A powerful supplier supplies a product important as an input to the individual firms in the industry. The more important the supplier's product or service is to the industry, the more power the supplier can exert by raising prices. This decreases profits for the individual firms. Another example of a powerful supplier is when the sales volume of the supplier's product or service is much larger than the purchasing volume of the individual firms in the industry. In this case, the supplier is said to be more concentrated than the industry it sells to, and the supplier is in a strong position to raise prices.

3.5. Intensity of Rivalry Among Competitors

The final competitive force is rivalry among existing firms. Tactics used can either increase or decrease profits for the industry. For instance, price competition tends to decrease profits. On the other hand, increased advertising may increase demand or increase product differentiation for the benefit of all firms in the industry.

Two reasons for increased rivalry among firms are the existence of numerous or equally balanced competitors and the existence of high exit barriers. Exit barriers are barriers to leaving the industry. For example, a firm may use highly specialized equipment for its work. In order to exit the business, the firm must sell the specialized equipment to someone who wants to use it in the same business. If a buyer cannot be found, the value of the specialized equipment is greatly reduced and often the equipment must be scrapped. Thus there are high costs associated with exiting the business.

The concept of exit barriers and entry barriers is a very important determinant of profits. With high exit barriers, returns are risky because of a high degree of rivalry among competitors. Unsuccessful firms will stay and fight it out with the rest of the industry. With low entry barriers, the potential profits are reduced as mentioned previously. The industries with the best potential for profitability are those in which the entry barriers are high and the exit barriers are low. Then entry will be deterred while unsuccessful, competing firms will be able to leave the industry. This leads to high and stable profits.

The implications for the profits of an individual firm within the industry stem from the fact that some firms consistently outperform others. An individual firm's profits are determined by its competitive position within the industry. A low degree of competition in the industry helps a firm obtain a strong position within the industry. The number of firms in the industry, economies of scale, and costs of entry and exit all influence the degree of competition. Finally, a key to a firm's competitive position is its ability to implement. A firm is the most profitable if it is in a favorable industry, has a strong competitive position in its industry, and can execute quickly to meet the challenges of changing business environments.

Chapter 5

GENERIC STRATEGIES

Whether an organization is large or small, the general process of strategic planning is similar. Formulating a competitive strategy involves a process of matching a firm's strengths to opportunities associated with its external business environment. Particular attention should be placed on what a firm does well in relation to other firms. Strategic planning is the way to begin finding answers to these concerns.

First, both large and small firms should analyze the business environment that is common to all firms in the industry. Executives assess the strength of competitive forces as described in the previous chapter. This will determine the potential for long-term profitability in the industry.

Next, a firm should analyze the strategies employed by other firms within the industry. The firm must identify its own strengths and weaknesses relative to competitors.

Finally, a choice can be made as to which strategy the firm should pursue. The strategy needs to be consistent with the organization's goals, capabilities, and business environment in order to be believable and realistic. Firms can choose strategies already employed by other firms in the industry or create a new strategy. The firm's strategy should optimize the expected profits relative to the expected costs and risks associated with that strategy.

While the process of strategic planning is similar for large and small organizations, the best generic strategies for each generally are not. The strategies are shown in Table 5.1. Large organizations are usually the most successful with strategies of overall cost leadership or

Table 5.1. Competitive Strategies to Outperform Other Firms in an Industry[8]
1. Cost leadership 2. Differentiation 3. Focus

industry-wide differentiation. Small organizations are more successful at focusing on a particular segment of the market.

1. COST LEADERSHIP

The cost leader is the firm that can produce the product or service with the lowest cost. Overall cost leadership is a dominant theme in the media, but there is not much room for cost leaders. Cost leadership often requires high relative market share, aggressive use of economies of scale, and use of the learning curve to reduce costs. These tactics generally require a large organization. The payoff is that a firm's cost position allows it to obtain defenses against the five competitive forces and realize above average returns. Such a firm has protection from rivalry because its lowest costs result in profits even after rivals lower their prices. There are barriers to entry because of economies of scale, and the firm is usually in an advantageous position compared to substitutes. Finally, there is bargaining power with buyers and suppliers because bargaining can erode profits only until all but the most efficient firm is eliminated.

2. DIFFERENTIATION

Differentiation involves creating a product or service that is viewed industry-wide as unique. The stipulation that the uniqueness be industry-wide implies a large market share is required. Thus large firms are the best suited to employ this strategy. The differentiation can be

---

8.   Michael E. Porter, *Competitive Strategy: Techniques for Analyzing Industries and Competitors,* New York: The Free Press, 1980.

achieved by technology, performance, customer service, or brand loyalty to name a few. An example of differentiation has been Apple Computer with its line of Macintosh computers introduced in 1984. For years these personal computers were easier to use than the IBM and IBM-compatible personal computers. Apple Computer was able to charge a premium for its Macintosh computers for most of the 1980s and into the early 1990s as a result of differentiation due to ease of use.

As with cost leadership, the differentiation strategy provides protection from the five competitive forces but in a different manner. For example, brand loyalty means barriers to entry, protection from substitutes, and protection from rivalry because loyal customers are willing to pay more for the advantages of a differentiated product. Differentiation leads to high margins which gives the firm some power in bargaining with suppliers. There is also some power in bargaining with buyers because they lack alternatives by definition of a differentiated product.

3. FOCUS

The generic strategy for small organizations is the focus strategy. Here a particular buyer segment is targeted. This is different than the strategies of cost leadership and differentiation which consider an entire industry. Within a market niche, a small organization can achieve cost leadership or differentiation. Because a small organization can stay focused on serving a particular market segment, it can sometimes serve this target market more effectively and efficiently than a large organization. As mentioned earlier, differentiation and cost leadership provide protection against the five competitive forces. This is true both in cases of an entire industry or a focused, target market. In addition to these defenses, focus can lower threats of substitutes and competition from rivals by selecting market segments where these competitive forces are weakest.

The outcome of strategic planning is the development of a course of action to achieve success. The organization has a clear picture of what it wants to achieve and how to accomplish it. The strategic plan defines the most important tasks to be successful.

As an engineer, you need to understand this plan and the strategy your organization is pursuing. If your organization is pursuing a cost leadership strategy, activities to reduce costs are the primary concern.

If differentiation is the strategy, you need to understand the performance and features of competitors' products. This understanding will help you determine how your products can be differentiated from the competitors' products. And if your organization is pursuing a focus strategy, get intimate knowledge of the target market. This knowledge will help you determine the features of a product or service that will be most beneficial to the customer. Determine how your activities can make a positive impact, and make it happen.

Chapter 6

STRATEGIC PLANNING CASE STUDY—MICROSOFT (1986)

To illustrate the strategic planning process, let's take a look at an example. This example gives you an idea of what might be going through the minds of executives as they formulate a plan. Chapter 7 discusses what engineers could have done once this plan was defined.

Microsoft has been one of the most successful technology companies over the last decade and is the largest personal computer software company. However back in 1986, it was the second largest personal computer software company behind Lotus Development Corporation. This example will show the strategic plan developed by Microsoft in its effort to overtake Lotus. First, the business environment of the software industry will be analyzed by determining the strength of the five competitive forces. Next the capabilities of Microsoft will be assessed, and finally its goals will be discussed.

1. BUSINESS ENVIRONMENT: THE SOFTWARE INDUSTRY IN 1986

Since the late 1970s, software firms had mushrooming demand for their products due to the phenomenal debut of the personal computer.[9] Personal computers gained much credibility with the entrance of IBM to the market in 1981. Throughout the early 1980s, rapidly changing hardware technology was a main factor for the rapid growth in software

---

9.   Tom Graves, "Computers & Office Equipment Current Analysis," *Standard & Poor's Industry Surveys,* Jan. 30, 1986, C61.

demand. However software firms, like the hardware firms before them, discovered that growth rates of greater than 40% annually were difficult to sustain over the long-term. This was partly because high levels of demand and profitability attracted new competitors.

1.1. Threat of New Entrants

In the late 1970s and early 1980s, the software industry was characterized as having low entry barriers.[10] Unlike almost any other industry, a person working alone could create products that competed with the industry leaders. However with the increasing number of software firms and with the shift towards the corporate market, fixed costs were increasing. Software firms needed to increase expenditures for advertising and distributing its products, and for research and development to develop new products. Such costs remained fixed whether a firm sold one copy of a software program or a million copies. These higher fixed costs increased the risks for software firms and raised the barriers to enter the software industry.

1.2. Threat of Substitution

The main threat of substitution came as a result of rapidly changing technology. As new computer hardware was developed, new software programs were created making old software obsolete. This led to short product life cycles and increased risk for the software firms. Software firms needed to be innovative with new software programs in order to survive because of the rapid advances in hardware technology. This trend would undoubtedly continue in the future as the capability of personal computers approached the capabilities of minicomputers and mainframe computers.

However, there were certain barriers to substitution for some of the best selling, business software packages. Some of these packages showed remarkable stability partly because of limited retail shelf space. Another factor was that once users had spent hours learning to use a particular package, they were reluctant to switch to another package unless the advantages to change were obvious.

---

10.   Robert J. Martorana, and Harvey Katz, "Computer Software & Services Industry," *The Value Line Investment Survey,* Sept. 19, 1986, 2112.

1.3. Bargaining Power of Buyers

In the early 1980s, the bargaining power of buyers was not particularly strong. As hardware prices came down and personal computers became increasingly indistinguishable, the personal computer became more of a commodity product. In this environment, software became the primary means of adding value to customers' computer systems.

The value of software to the customer was expected to continue in the future, but the large corporate market segment was beginning to gain bargaining power as of 1986. There was growing corporate dependence on personal computers. Large corporations were fast incorporating personal computers into the mainstream of their data processing operations.[11] John Imlay, president of Management Science America (MSA), predicted in 1985, "While 90% of corporate computing is now done on mainframes, 90% will be done on personal computers by 1995."[12] Corporate customers were calling for increasingly innovative software products and increasingly accommodative purchase terms.

Corporate customers were looking for software companies who could provide broad-based solutions that required the choice and integration of many hardware, software, and networking options. This helped accelerate the merger activity of software firms which began in the mid-1980s.

Also begun in the mid-1980s was the negotiation for site licensing. Site licensing allowed a company to make a specified number of copies of a software package for a fee. This helped lower the average price of software products for corporate customers.

1.4. Bargaining Power of Suppliers

Suppliers to the software industry could be broken down into three main areas: technology, manpower, and materials. These suppliers had varying degrees of bargaining power.

---

11. Jan Snyders, "Software: Then, Now and in the Future," *Infosystems,* May 1984, 44.

12. Fred V. Guterl, "PC Software Firms Seek New Growth," *Dun's Business Monthly,* Nov. 1985, 68.

1.4.1. Technology

As stated earlier, rapid advances in hardware technology fueled the rapid growth in the software industry. Future growth in software was highly dependent on the continued advances in hardware technology. It was imperative for software firms to understand new hardware advances and to quickly introduce innovative software products. Fortunately for the software industry, increasing incorporation of hardware standards made personal computers more of a commodity product. Thus learning about hardware advances became a less formidable task. The bargaining power that individual hardware firms had over software firms was decreasing.

1.4.2. Manpower

The rapid growth in the software industry drove up the demand for computer science professionals and engineers. There was high demand for software engineers in 1986. Fortunately for the software firms, computer science and related majors became a popular field of study, so an adequate supply of people were continually entering the work force.

1.4.3. Materials

The main materials for the software industry were computer hardware and magnetic disks. As mentioned earlier, the bargaining power of the computer suppliers was decreasing. The magnetic disks were in ample supply from many sources. The disk suppliers were not able to exert much price pressure because of the ease in switching to alternate suppliers.

1.5. Intensity of Rivalry Among Competitors

The rivalry between software firms was beginning to intensify because of the staggering number of programs available. In 1984 it was estimated there were 20,000 software programs on the market for about 5 million personal computers in use.[13] In the major categories, there were at least 200 word processing packages, 150 spreadsheets, 200 database packages, and 95 integrated software packages. The

---

13.   "The Shakeout in Software: It's Already Here," *Business Week*, Aug. 20, 1984, 102.

competition was intensifying because consumers tended to buy only one version of each major application type.

With more competition, software vendors needed to spend more on advertising to get customer attention and win retail shelf space. For example, Ashton-Tate spent $10 million in 1984 on an advertising campaign to back its new Framework integrated package. The environment was tending to favor larger software firms as the industry matured.

In summary, the software industry growth rate for the late 1980s was expected to slow down compared to the phenomenal growth rate of 40% seen in the early 1980s. Reasons for the slowdown included higher fixed costs for introducing new products and price pressure for existing products.

However the outlook was far from bleak. The software industry was expected to grow at an annual rate of 20%-30% for the rest of the 1980s. This growth was projected to be fueled by advances in hardware technology, demand for easier to use software, and the creation of technological standards to enable easier networking between computers.

In this environment of 1986, it became increasingly important for software firms to produce innovative products and to have marketing clout. The challenge was to provide a growing corporate market segment with innovative products that were easy to use. Also as the costs of product development and product advertising increased, smaller software firms were increasingly likely to seek shelter through mergers with the major companies.

2. CAPABILITIES: MICROSOFT IN 1986

Microsoft Corporation was founded in 1975 with the introduction of its BASIC language interpreter.[14] Its mission was to bring the benefits of personal computer technology to the greatest number of people. The company had been enormously successful to this point. Microsoft's sales in 1986 were $197.5 million which put Microsoft

---

14.   Microsoft Corporation Form 10-K, *Securities and Exchange Commission,* for the fiscal year ended June 30, 1986.

second behind industry leader Lotus Development Corporation.[15] Microsoft's two main product lines were systems software, including operating systems and language products, and applications software.

In 1986, Microsoft employed 1,153 people, 960 domestically and 193 internationally.[16] Of this total, 362 were in product development and 460 were in sales and marketing.

The capabilities of Microsoft will now be examined from three standpoints: finance, marketing, and product development.

2.1. Finance

Microsoft was second in the industry in terms of sales behind Lotus Development Corporation. Table 6.1 shows the sales and profits for the two companies through 1986.

Both companies had been growing tremendously. Lotus Development's sales growth averaged 75% per year from 1983 to 1986. Microsoft's sales growth averaged 58% over the same period, and its earnings soared 63% in 1986.

Table 6.1. Microsoft and Lotus Development Sales and Profits for 1983 through 1986

Company		1983	1984	1985	1986
Lotus[17]	Sales (millions)	$53.0	$157.0	$225.5	$282.9
	Net Profit (millions)	$13.7	$36.1	$38.2	$48.3
Microsoft[18]	Sales (millions)	$50.1	$97.5	$140.4	$197.5
	Net Profit (millions)	$6.5	$15.9	$24.1	$39.3

15. Richard Brandt, "Microsoft Moves into the Passing Lane," *Business Week,* Aug. 18, 1986, 102.

16. See note 14.

17. *The Value Line Investment Survey,* June 11, 1993, 2124.

18. *The Value Line Investment Survey,* June 11, 1993, 2125.

Microsoft had $10 million in cash on hand in 1985,[19] and its initial public offering in 1986 raised another $39 million. Cash-rich and debt-free, the company was poised to grab market share from industry leader Lotus.

2.2. Marketing

2.2.1. Products

Microsoft was founded by William Gates, a programming whiz who dropped out of Harvard to start the company, and his friend Paul Allen.[20] The company licensed MS-DOS to IBM in 1981 as the basic operating system for its Personal Computer. Since then, MS-DOS was the standard for IBM's PC and all PC-compatible machines. Thus in 1986, Microsoft was already the undisputed leader in selling operating system software. Its systems software products, including operating systems and computer language products, accounted for 53% of the company's sales.[21]

However, partly because it entered the market late, Microsoft had never achieved the same dominance in applications software. This accounted for 37% of the company's sales.[22] Prior to 1986, Microsoft was most successful in applications by avoiding major competitors. Its one program to compete directly with a big rival was its spreadsheet called Multiplan. It was never a hit on the scale of Lotus' 1-2-3, the industry's best selling applications program. Also, Microsoft had avoided the market for database management software controlled by Ashton-Tate's dBase programs. It had concentrated on such areas as word processing where there was no dominant program. Microsoft Word was one of the best selling programs, and it had sold over 250,000 units as of 1986.

Microsoft had avoided the dominance of Lotus and Ashton-Tate in applications software by taking a chance on Apple's Macintosh

---

19. Jonathan B. Levine, "Microsoft: Recovering From Its Stumble Over Windows," *Business Week,* July 22, 1985, 107.

20. Bro Uttal, "Inside the Deal That Made Bill Gates $350,000,000," *Fortune,* July 21, 1986, 23.

21. Microsoft Corporation Form 10-K, *Securities and Exchange Commission,* for the fiscal year ended June 30, 1986.

22. See note 21.

computer. Microsoft sold 50% of all Macintosh programs, making it the leading supplier of Macintosh software. In October, 1985, Microsoft introduced Excel, a spreadsheet integrated with database and business graphics modules for use with the Macintosh. It won rave reviews and was the best selling Macintosh program in 1986.

One of the strategies Microsoft had to attain dominance in applications software rested with the introduction of its Windows program for the IBM PC in November, 1985. Windows was a cross between an operating system and an applications program. The program simplified the commands needed to run a computer. Instead of keystroke commands, computer operators needed only to point an electronic mouse at symbols. Windows could also display several programs on the computer screen at once and shift data back and forth among them.

Windows provided a foundation for the next generation of graphics-based applications. These new applications used drop-down menus, icons, and mixed text and graphics to give the user a more intuitive way to interact with the computer. It was an innovation in software technology that made computers easier to use.

As of 1986, Microsoft had sold 4,000 development kits to let software writers from other companies develop applications for Windows and make Windows an industry standard.

2.2.2. Distribution

One of Microsoft's strengths was its established and accepted distribution channels. These included marketing products both domestically and internationally as well as working directly with large corporate customers.

Domestic OEM distribution

Microsoft's operating systems were marketed primarily to Original Equipment Manufacturers (OEMs), giving the OEMs the right to copy and distribute Microsoft's operating systems with the OEMs' microcomputers. Microsoft had OEM agreements with virtually all of the major domestic microcomputer OEMs including IBM, Tandy, and Compaq.

Domestic retail distribution

This was the primary channel for Microsoft's applications software. The company marketed its products in the retail channel through independent distributors and dealers, large volume dealers such as Computerland and Businessland, and dealers who emphasized large business customers. In recognition of the importance of obtaining large orders from corporate customers, Microsoft established a group to market applications products directly to those customers.

International OEM distribution

Microsoft distributed to and maintained an active business and technical information relationship with a number of Japanese microcomputer manufacturers including NEC, Mitsubishi, and Matsushita. They also had OEM marketing and business relations with European OEMs including Siemens, Philips, and Ericsson.

International retail distribution

Microsoft had established marketing, distribution, and support subsidiaries in Canada, the United Kingdom, Germany, France, Japan, Italy, and Australia. It had a practice of localizing its retail products, including translating user messages and documentation, into the language of each respective country.

2.3. Product Development

The personal computer software industry was characterized by rapid technological change. To be successful required a continuous, high level of expenditures for the enhancement of existing products and the development of new products. As of July, 1986, Microsoft employed over 300 people engaged full time in software development. During fiscal 1984, 1985, and 1986, the company spent $10.7 million, $17.1 million, and $20.5 million, respectively, on product development and enhancement activities.[23] This represented 11%, 12%, and 10%, respectively, of net revenues for these years.

---

23. Microsoft Corporation Form 10-K, *Securities and Exchange Commission,* for the fiscal year ended June 30, 1986.

Most of Microsoft's software products were developed internally which allowed them to maintain close technical control. A crucial factor in the success of a new product was getting to market quickly to respond to a new user need or an advance in hardware technology. In this regard, Microsoft could take advantage of being the major player in operating system software as a result of its development of the MS-DOS system used by IBM and other hardware vendors. A 1985 agreement between Microsoft and IBM for future operating system development helped Microsoft continue its leadership in this area.

However even with this advantage, the company had some problems developing its products quickly. William Gates, Microsoft's co-founder, was noted for his haphazard style of management. Microsoft's innovative Windows program was over a year behind schedule. Their problem was it outgrew its small-time style faster than Gates could handle. "[Microsoft's] founders had great intuition, but they could only [manage] so many products," said Gates.[24]

In the rapid growth that followed the enormous popularity of MS-DOS, Gates tried to take personal charge of five product lines. As a result, little attention was paid to tailoring programs to customers' needs. Key planning decisions were often delayed or not made. As an example, it took management more than one year to realize Windows required too much computer memory for the commercial market.[25] There was a lack of focus and continuity as new deals were cut to develop software for major customers and yanking developers off other projects to work on new ones.

To his credit, Gates recognized the problems and took himself out of much of the day-to-day operations. He avoided the fate of his good friend Steven Jobs, who was forced out of operations at Apple Computer in a highly publicized move in June, 1985. Gates brought in an outsider as president in 1982, and then later hired Jon Shirley away from Tandy Corporation.

In August, 1984, Microsoft took serious action by reorganizing around two divisions: systems software and business applications. Each division could then synchronize product development and market

---

24.   Jonathan B. Levine, "Microsoft: Recovering From Its Stumble Over Windows," *Business Week,* July 22, 1985, 107.

25.   See note 24.

introductions. They also started holding product review meetings on a regular basis. Confident in its new organization, Microsoft made an early announcement of its Excel spreadsheet for Macintosh computers in May, 1985 when shipments were scheduled for September, 1985.

3. GOALS

Microsoft's main goal was to overtake Lotus Development Corporation as the No. 1 independent software supplier by increasing its market share in applications software. A major obstacle to increase applications market share was brand loyalty. Corporate customers had spent several years training employees to use Lotus 1-2-3 and Ashton-Tate's dBase programs. The generic strategy for gaining applications market share was product differentiation. Microsoft had to differentiate its applications programs as easier to use and more powerful than the competition's products.

An area of opportunity Microsoft saw to increase its share of the applications market was introducing a new version of MS-DOS. This would then require a new generation of applications programs for the higher powered personal computers hitting the market. Originally in 1981, MS-DOS was designed to work with the Intel 8088 microprocessor that was used in the IBM PC. A new version of MS-DOS was needed to take full advantage of Intel's 80286 microprocessor used in IBM's new PC/AT and compatibles. With the new version of the operating system, all software companies would have to develop new versions of their applications programs to take advantage of the computer system's capabilities. Microsoft planned to introduce its new version of MS-DOS in early 1987.

Once the new MS-DOS was introduced, Microsoft's big chance was to bring out a PC version of its Excel spreadsheet to compete directly with Lotus 1-2-3.

The other part of Microsoft's strategy to overtake Lotus was to make Windows an industry standard. Once it became a standard, Microsoft hoped to exploit its intimate knowledge of the system to develop the best applications for it. As early as 1986, nearly half of Microsoft's development staff was working on applications for Windows.

To make Windows an industry standard, Microsoft was counting on other software companies to write a broad range of applications to go with Windows. However there were many stumbling blocks to face before standardizing Windows. The key rested with IBM's endorsement. IBM was rumored to be developing a rival product, called TopView, which would employ easy-to-use graphic symbols.

Other stumbling blocks included the delay of more than a year in introducing Windows. This created uncertainty among the other software companies. Also major software companies were reluctant to write applications for Windows because they did not want to push demand for a Microsoft product. For example, as of 1986 Lotus had agreed to write some programs for Windows but did not plan to write a new version of 1-2-3 for it. Lotus 1-2-3 remained only as an MS-DOS application at that time.

In summary, Microsoft's goal was to become the leading independent software supplier. To achieve this, the company took into account the business environment for the software industry and evaluated its own capabilities relative to their competition. Microsoft decided to pursue three main tasks.

1) **Maintain its dominance in personal computer operating systems.**
   Microsoft would release a new version of MS-DOS in early 1987 to take advantage of Intel's new 80286 microprocessor used in the IBM PC/AT and compatibles.

2) **Release a version of its Excel spreadsheet for the IBM PC to compete directly with Lotus 1-2-3.**
   With a more powerful microprocessor and a more powerful operating system, computer users could upgrade to more powerful software. Microsoft could exploit its intimate knowledge of the operating system to develop the best applications for it.

3) **Make Windows an industry standard.**
   The key rested with IBM's endorsement. Microsoft planned to win this endorsement by trying to get as many independent software companies as possible to write applications for Windows.

The result of following this strategic plan will be covered in Part III on Finance.

Chapter 7

HOW TO CONTRIBUTE AS AN ENGINEER

This chapter discusses what engineers can do to maximize their benefit to the organization. The first step is to understand the strategic plan for the company. Start by asking your manager about it. He or she may suggest talking to other people to gain other perspectives. Gain an understanding of the business environment, company capabilities, and company goals.

Microsoft pursued a differentiation strategy. It wanted to establish its applications and systems software as easy to use with features customers wanted. Areas for you to consider with a differentiation strategy are: (1) the features customers want, (2) how your organization's products can be differentiated from competitors' products, and (3) how new products can be developed more quickly. The main responsibility for these areas may rest with other people, but engineers have a unique perspective and can make suggestions about how to improve in these areas.

1. KNOWING WHAT CUSTOMERS WANT

Start by talking with your manager and people from Marketing and Sales. Find out what they know about what customers want. Volunteer to talk with customers directly if the opportunity arises. Sometimes Marketing conducts customer surveys or customer visits to find out such information, and it is essential to have technical people involved in such activities. This may be an opportunity for you to get involved. Document information you gather so it can benefit your organization.

If there isn't such an opportunity, understand the information gathered from ongoing customer surveys and visits.

2. HOW YOUR PRODUCTS COMPARE WITH COMPETITORS

Again start by talking with your manager and people from Marketing and Sales. There may already exist a lot of information about how your company's products compare with competitors' products. If there is such information, understand the desirable features of the competitors' products which can be improved. Understand areas where competitors' products are weak which can be opportunities for differentiation by your company. If such information does not exist, talk to your manager about how such information can be obtained. There may be opportunities for you to get involved with analyzing competitors' products. Remember to document information you gather. Executives are very interested in information about competitors, and this is one way to gain more visibility in your current position and to get a better understanding of product features your organization should work on.

Competitor analysis is an area which must have involvement from engineers. Marketing and Sales personnel many times do not have the technical depth you can provide. But Marketing and Sales must be relied upon because they know who are the competitors and the customers for your products. Thus the most effective competitor analysis programs require close teamwork between people in Marketing and Sales and people in Engineering.

3. THE PRODUCT DEVELOPMENT PROCESS

Getting new products to market quickly was certainly an important aspect for Microsoft in 1986. Engineers working in product development have a unique perspective regarding the product development process and the resources allocated for product development because they are intimately involved with the process.

If you are in such a situation, one of the ways you can contribute to the organization is to give feedback and suggestions to your manager of how the product development process can be improved. As you go through aspects of your product development work, jot down ideas of how certain tasks can be done better (i.e. faster, cheaper, or smarter). Use methods learned from Part I when presenting these ideas to your manager. Remain objective and stay focused on the task of improving the product development process. Do not blame other people or get emotional. Focus on the situation at hand.

Getting products to market is frequently an important aspect for engineering work. Methods for improving your work in this area are covered in Part IV on Statistics and Part V on Project Management.

Chapter 8

PITFALLS OF STRATEGIC PLANNING

The value of strategic planning is not realized if the planning process or the plan is misunderstood. The value of planning rests with the process of planning itself. This process entails the executives analyzing the industry's business environment, assessing the organization's capabilities, considering options, picking a strategy, and implementing it. The executives should communicate the plan to the work force, so everyone in the organization understands it and works toward common goals. This is also your opportunity to provide feedback. After this process, the planners should know what explicit assumptions were made and how to monitor the organization's progress. The key factors to be managed are known.

The plan is not a definitive statement about future events. It should not be relied upon as a forecast or a prediction of the future. There is a growing tendency to do so as the world becomes more uncertain, but this is not the intent of a strategic plan. The future should be viewed in relation to the assumptions made in the plan.

The plan needs to be flexible. It should include a contingency plan which is adopted when certain assumptions become invalid. For the executives, planning is a continual process of monitoring the benchmarks and the specified assumptions.

A final point is the tendency to include quantitative measures (such as dollar and time goals) in the plan because they can be communicated without ambiguity. This leads to a short-term focus and diverts attention away from the underlying forces that change the environment. The plan should emphasize events and their assumptions rather than show only financial aspects.

KEY POINTS: Strategic Planning

1) The main purpose of a strategic plan is to define the most important tasks for an organization to achieve long-term success. One of the most important responsibilities of the executives is to develop a strategic plan. This is done by analyzing the business environment, the capabilities of the organization, and the goals of the organization.

2) The business environment warrants the most attention because it usually cannot be controlled as easily as goals and capabilities. The environment must be analyzed by determining the strength of the following competitive forces:

   - Threat of new entrants
   - Threat of substitution
   - Bargaining power of buyers
   - Bargaining power of suppliers
   - Intensity of rivalry among competitors

3) There are three generic strategies. Large organizations are usually the most successful with strategies of overall cost leadership or industry-wide differentiation. Small organizations are more successful at focusing on a particular market segment.

4) The strategic plan is not a definitive statement about future course of events. The plan must be constantly updated by assessing events versus assumptions made in the plan. The plan should balance long-term, qualitative aspects with short-term, quantitative aspects.

5) As an engineer, you must understand the strategic plan. The strategic plan defines the most important activities for the success of your organization. Thus it defines where you should focus your efforts in order to maximize your benefit to the organization.

6) Know what generic strategy (or strategies) your organization is pursuing. Ways that you can contribute significantly to the organization differ depending on the strategy.

   For cost leadership strategy:

   - Find ways where costs can be reduced.

For differentiation or focus strategies:

- Understand what your customers want.
- Know how your products compare with competitors' products.
- Suggest ways the product development process can be improved.

SUGGESTED ACTIVITIES: Strategic Planning

1) Understand the strategic plan for your company and your organization. Start by asking your manager. Know your organization's goals, capabilities, and business environment. Identify any assumptions made by the executives. Know the most important tasks for your organization to achieve long-term success.

2) Determine the needs of the customers who buy your organization's products or services, and understand how they compare with competitors' products. Begin this activity by talking with personnel in the Sales and Marketing department. Volunteer to talk directly with customers, and volunteer to analyze competitors' products.

3) Identify the generic strategy your organization is pursuing in order to achieve long-term, above average profitability. The generic strategies are cost leadership, differentiation, and focus strategies.

4) Based on the generic strategy and strategic plan for your organization, find ways to contribute towards achieving the plan. Talk with your manager and other managers in the organization to see how your activities align with tasks that are important for the organization.

REFERENCES: Strategic Planning

Technology

Boar, B. H., *The Art of Strategic Planning for Information Technology: Crafting Strategy for the 90's*, New York: John Wiley & Sons, Incorporated, 1993.

Burgelman, R. A. and Maidique, M. A., *Strategic Management of Technology, 2nd Edition*, Burr Ridge: Richard D. Irwin, 1995.

Concordia University Staff, Editor and McTavish, R., Editor, *Linking Marketing & Technology Strategies*, Chicago: American Marketing Association, 1990.

Dussauge, P., Hart, S., and Ramanantsoa, B., *Strategic Technology Management*, New York: John Wiley & Sons, Incorporated, 1992.

Loveridge, R., Editor and Pitt, M., Editor, *The Strategic Management of Technological Innovation*, New York: John Wiley & Sons, Incorporated, 1992.

Madu, C. N., *Strategic Planning in Technology Transfer to Less Developed Countries*, Westport: Greenwood Publishing Group, Incorporated, 1992.

Male, S. P. and Stocks, R. K., *Competitive Advantage in Construction*, Newton: Butterworth-Heinemann, 1992.

McGrath, M. E., *Product Strategy for High-Technology Companies: How to Achieve Growth, Competitive Advantage, & Increased Profits*, Burr Ridge: Irwin Professional Publishing, 1994.

Utterback, J. M., *Mastering the Dynamics of Innovation: How Companies Can Seize Opportunities in the Face of Technological Change*, Boston: Harvard Business School Press, 1994.

Zangwill, W. I., *Lightning Strategies for Innovation: How the World's Best Companies Create New Products*, New York: Free Press, 1992.

General

Bowman, C., *Essence of Strategic Management*, Englewood Cliffs: Prentice Hall, 1991.

Brown, S., *Strategic Manufacturing for Competitive Advantage: Transforming Operations from Shop Floor to Strategy*, Englewood Cliffs: Prentice Hall, 1995.

Cook, K. J., *AMA Complete Guide to Strategic Planning for Small Business*, Lincolnwood: N T C Publishing Group, 1994.

Cope, R., *High Involvement Strategic Planning: When People & Their Ideas Really Matter*, Oxford: Planning Forum, 1989.

Dixit, A. and Nalebuff, B. J., *Thinking Strategically: The Competitive Edge in Business, Politics, & Everyday Life*, New York: W. W. Norton & Company, Incorporated, 1993.

Finnie, W. C., *Hands-On Strategy: The Guide to Crafting Your Company's Future*, New York: John Wiley & Sons, Incorporated, 1994.

Judson, A. S., *Making Strategy Happen: Transforming Plans into Reality*, Oxford: Planning Forum, 1990.

Karger, D. W., *Strategic Planning & Management: The Key to Corporate Success*, New York: Marcel Dekker Incorporated, 1991.

Morden, A. R., *Business Strategy & Planning: A Strategic Management Approach*, New York: McGraw-Hill Companies, 1993.

Normann, R. and Ramirez, R., *Designing Interactive Strategy: From Value Chain to Value Constellation*, New York: John Wiley & Sons, Incorporated, 1994.

Pfeiffer, J. W., Goodstein, L., and Nolan, T., *Applied Strategic Planning: How to Develop a Plan That Really Works*, New York: McGraw-Hill Companies, 1993.

Porter, M. E., *Competitive Advantage: Creating & Sustaining Superior Performance*, New York: Free Press, 1985.

Porter, M. E., *Competitive Strategy: Techniques for Analyzing Industries & Competitors*, New York: Free Press, 1980.

Saunders, J. and Hooley, G., *Competitive Positioning: The Key to Marketing Strategy*, Englewood Cliffs: Prentice Hall, 1993.

Schwartz, P., *Art of the Long View: The Path to Strategic Insights for Yourself & Your Company*, New York: Doubleday & Company, Incorporated, 1991.

Sifonis, J. G. and Goldberg, B., *Dynamic Planning: The Art of Managing Beyond Tomorrow*, New York: Oxford University Press, Incorporated, 1994.

Stopford, J. M. and Baden-Fuller, C., *Rejuvenating the Mature Business: The Competitive Challenge*, Boston: Harvard Business School Press, 1994.

Taylor, B. and McNamee, P. B., *Developing Strategies for Competitive Advantage*, Elkins Park: Franklin Book Company, Incorporated, 1990.

Templeton, J. F., *Focus Group: A Strategic Guide to Organizing, Conducting & Analyzing the Focus Group Interview*, Burr Ridge: Probus Publishing Company, Incorporated, 1994.

Thomas, H., Editor, *Building the Strategically-Responsive Organization*, New York: John Wiley & Sons, Incorporated, 1994.

Watson, G. H., *Strategic Benchmarking: How to Rate Your Company's Performance Against the World's Best*, New York: John Wiley & Sons, Incorporated, 1993.

Nonprofit/Government

Burkhart, P. J. and Reuss, S., *Successful Strategic Planning: A Guide for Nonprofit Agencies & Organizations*, Thousand Oaks: Sage Publications, Incorporated, 1993.

Hay, R. D., *Strategic Management in Non-Profit Organizations: An Administrator's Handbook*, Westport: Greenwood Publishing Group, Incorporated, 1990.

Moore, M. H., *Creating Public Value: Strategic Management in Government*, Cambridge: Harvard University Press, 1995.

Pappas, A. T., *Reengineering Your Nonprofit Organization: A Guide to Strategic Transformation*, New York: John Wiley & Sons, Incorporated, 1995.

PART III

FINANCE

Understanding Your Organization's Scorecard

Part II on strategic planning emphasized that you must understand what the most important tasks are for the organization to achieve long-term success. The study of finance aids in this analysis by providing a framework for measuring success. The motivation for you to know the basics of finance is to comprehend this language of managers and executives. Thus you can show your impact to the organization in a manner they understand.

There have been problems with engineers attempting to involve themselves with finance.

1) **The business focus is short-term.**

Too often the perception an engineer has of finance is meeting the sales numbers for the month. Concentration on monthly sales leads to a short-term focus. It is important to realize finance can also be used to measure and plan ahead for long-term success. The purpose of Part III is to define how financial planning ties in with the strategic plan for the organization.

2) **Finance is "dry" because it reports results but does not relate well to people.**

When speaking of finance, it is important to focus on the human side—what engineers can do to affect changes. Then the numbers have meaning and use.

Engineers contribute most significantly to a firm when they can achieve real time results towards making the organization a long-term success. Part II of the book showed how the strategic plan defines the most important tasks. Part III defines what the key financial indices are for determining and measuring success. It will be shown that the key financial indices depend on the growth goals for the firm. By understanding the most important tasks and the key financial indices, you can align yourself towards making the maximum positive impact.

Part III consists of Chapters 9 through 15. Chapter 9 describes a generic model for how a business functions. It shows how the organization is a dynamic system encompassing both operations and investment decisions and highlights key financial measures for the system. Chapter 10 explains balance sheets and income statements. That will be followed by a chapter on how to analyze the finances of a business. This will include methods to determine financial health and a method for overall analysis called financial linkage analysis. Financial linkage analysis shows how key measures are tied together to determine the sustainable rate of growth for the business. Included will be two case studies on financial analysis. The first case study, Chapter 12, compares the finances of two disk drive companies, Quantum Corporation and Micropolis Corporation. The second case study, Chapter 13, revisits two software companies, Microsoft Corporation and Lotus Development Corporation. The financial results of implementing Microsoft's strategic plan (presented in Part II) will be studied. Chapter 14 shows how you can obtain financial information for your situation and how you can make a significant, positive impact on these finances. It stresses how productivity can be improved and where to focus your efforts depending on the growth strategy for the business. Chapter 15 pertains to those of you working for international companies or have international customers. It covers exchange rates in the foreign currency markets and what implications they have for exporters and importers.

Chapter 9

THE BUSINESS AS A DYNAMIC SYSTEM

To understand how different areas of the business fit together, it is helpful to view the business as one dynamic system (see Figure 9.1). The business can be thought of as two main areas upon which decisions need to be made in order to achieve long-term success. These areas are the operations area and the investment area.

1. OPERATIONS

The operations area is the area most familiar to engineers. As shown in Figure 9.1, operations involves the coordination of price, volume, and costs in order to achieve an operating profit. Let's look at the main concerns of the operations area.

First, the operations area must define the target market for the firm's products or services. This definition comes from the strategic plan, as discussed in Part II. Engineers are heavily involved in the next two concerns: deployment of assets and cost effectiveness. Deployment of assets can cover anything from the operation of the plant and equipment to how cash is used. The key measure in the operations area is the rate of return these assets produce, called return on assets. Return on assets is the percentage ratio of profits divided by assets. Details of this key measure and others will be reviewed in Chapter 10 on financial reporting. A glossary of financial reporting terms is shown in Table 10.1. A summary of the financial measure formulas is shown in Table 11.1.

Figure 9.1. The Business as a Dynamic System[26]

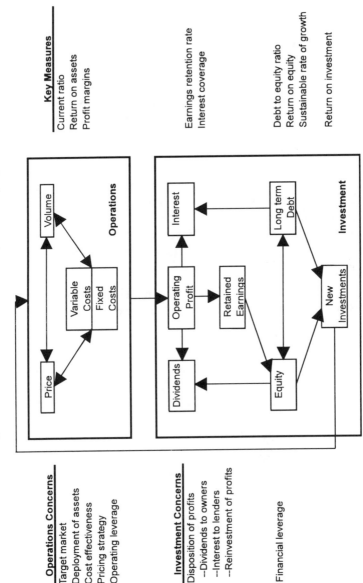

Cost effectiveness is how successfully the product or service can be engineered to lower its cost while maintaining the quality and performance required by customers of the target market. Costs can be broken down into variable and fixed costs. Variable costs are a function of the volume produced, but fixed costs are not a function of the volume produced. Categorizing costs into variable and fixed costs leads to the concepts of break even point and operating leverage.

Closely tied to the deployment of assets and cost effectiveness is the pricing strategy. As seen in this simplified model, price, volume, and costs are interdependent within the operations of the firm. A key measure for the operations is profit margin —the percentage of profits relative to sales. Another key measure for the operations is the current ratio, the ratio of current assets to current liabilities. This is a measure of the health of the operations area. Current assets are those assets that are cash or expected to be converted into cash within a year, and current liabilities are those debts due within one year. In other words, the current assets are used to pay off the current liabilities, and the current ratio shows how much cushion there is for the operations to pay off its short-term debts.

To aid in the analysis of the operations, it is important to understand the principles of the break even point and operating leverage.

1.1. Break Even Point

The concept of separating costs into fixed and variable costs leads to the basis of break even point analysis. The break even point is the point where there are no profits or losses. Operating at volumes above the break even point results in profits, and operating at volumes below the break even point results in losses. Equation (9.1) states profit, Y, as a function of volume, V. The components in this equation are unit price, P, unit variable cost, C, and fixed costs, F.

$$Y = VP - (VC + F) \tag{9.1}$$

This equation can be rewritten as:

$$Y = V(P - C) - F \tag{9.2}$$

26. Adapted from Erich A. Helfert, *Techniques of Financial Analysis,* Homewood, IL: Richard D. Irwin, Inc., 1982.

which shows that profit depends on the volume of goods or services sold times the difference between the unit price and the unit variable cost. This difference in unit price and variable cost is the contribution which offsets the fixed cost. To solve for the break even point, set $Y = 0$ in Equation (9.2) to get:

$$V_{break\ even} = \frac{F}{P - C} \qquad (9.3)$$

Examine Figure 9.2 as an example. Company A has fixed costs of $250,000. The price for its product is $1000/unit, and its variable costs are $500/unit. Thus for every unit sold, $500 (or $P - C$) gets contributed towards offsetting the fixed costs. By using Equation (9.3) to solve for the break even point, Company A breaks even by producing and selling 500 units.

Figure 9.2. Break Even Point Example

Company A

Fixed costs,	$F = \$250,000$
Unit price,	$P = \$1000$
Unit variable cost,	$C = \$500$

$$V_{break\ even} = \frac{F}{P - C} = \frac{\$250,000}{\$1000 - \$500} = 500\ units$$

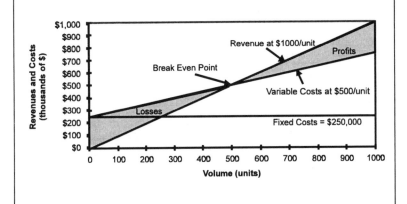

1.2. Operating Leverage

Introducing the concept of fixed and variable costs leads to the concept of operating leverage. Once the fixed costs have been recovered by the break even volume of operations, profits rise proportionately faster than the rate of increase in additional volume. The converse is also true. If the operations is operating at volumes below the break even point, the losses mount proportionately faster than the rate of decrease in volume.

This is illustrated in Figure 9.2. The shaded area at volumes above the break even point corresponds to profits, and the shaded area at volumes below the break even point corresponds to losses. Let's examine the various volume and profit/loss scenarios shown in Table 9.1. Equation (9.2) was used to calculate the profits and losses.

A series of 20% increases in volume above the break even point result in much larger than 20% jumps in profit growth. Similarly, a

Table 9.1. Company A
Profits and Losses vs. Volume

Company A

Fixed costs,	$F = \$250,000$
Unit price,	$P = \$1000$
Unit variable cost,	$C = \$500$

$$Y = V(P - C) - F = V(\$1000 - \$500) - \$250,000$$

Volume, V		Profits or (Losses), Y	
Units	% change	$	% change
500	----	$0	----
600	+20%	$50,000	∞
720	+20%	$110,000	+120%
864	+20%	$182,000	+65%
1037	+20%	$268,400	+47%

Volume, V		Profits or (Losses), Y	
Units	% change	$	% change
500	----	$0	----
400	-20%	($50,000)	∞
320	-20%	($90,000)	-80%
256	-20%	($122,000)	-36%
205	-20%	($147,500)	-21%

series of 20% decreases in volume below the break even point result in much larger than 20% increases in losses. However the relative change in profits and losses decrease the further away from the break even point.

In general, changes in operation close to the break even point, both above and below, are likely to produce sizable percentage swings in profits and losses. Changes in operations well above or below the break even point will cause lesser fluctuations.

In addition, the higher the relative fixed costs, the more powerful operating leverage becomes. Industries such as automobile manufacturing, steel, mining, forest products, construction, and semiconductor manufacturing require sizable investments in capital equipment. They are all subject to highly leveraged operations. A large portion of the production costs are fixed for a wide range of volumes. This tends to magnify profit swings as such companies operate around the break even point.

In contrast, service industries such as engineering consulting require much less investment in capital equipment, and thus have much lower relative fixed costs. The major cost in these companies is usually labor, and this cost can be adjusted by adjusting the number of employees as demand changes. Service industries are subject to much less profit swings caused by operating leverage.

Now that a simplified model for operations has been presented, it is important to understand how the operations and operating leverage can be influenced. There are three main elements which affect break even volume: fixed costs, variable costs, and price. The key driving force in this relationship depends on the relative magnitude of these elements in each particular organization. As shown in Equation (9.3), the break even point can be lowered, thus improving the profit position, by decreasing fixed costs, decreasing variable costs, or raising prices.

2. INVESTMENT

The investment area is concerned with dispositioning the operating profits. Hopefully some of the profits will be retained by the owners, and some will be reinvested back into the operations to fuel or sustain growth. The relative amounts of earnings retained by owners versus reinvestment into the operations depends on the financial strategy of the organization.

As shown in Figure 9.1, operating profits are dispositioned in three ways: dividends to owners, interest to lenders, and retained earnings. Each of these are influenced by past and current financial planning strategies. The rate of dividend payout affects the use of operating profits for reinvestment and growth. Interest payments are a matter of contractual obligations, and its significance depends on the relative amount of debt employed. The greater the proportion of debt, the higher the interest payments. Furthermore, high debt proportions generally require higher interest rates to be paid to compensate for the potential risk to the lenders. Retained earnings are what remains after dividends and interest payments are made. The key measures for the disposition of profits are the earnings retention rate and the interest coverage.

Figure 9.1 shows that the retained earnings are added to owners' equity. The retained earnings can be combined with new capital provided by investors (equity) and by lenders (long-term debt) to fuel growth by investing back into the operations. Key measures in this area include the ratio of debt to equity, the return on equity, the sustainable rate of growth, and the return on investment. New investment decisions should be based on the strategic plan and financial plan for the organization. New investments need to be consistent with the firm's operational characteristics, business objectives, and financial policies. A common tool to analyze and justify new investment is the return on investment method, and this will be described in Part V on Project Management. A concern in the investment area is how much debt to employ in order to increase profits. This involves the concept of financial leverage.

2.1. Financial Leverage

Financial leverage and operating leverage are similar in that both give the opportunity to benefit from fixed costs. Resources associated with the fixed costs can be employed to generate more profits. In the case of financial leverage, the advantage arises when funds borrowed at a fixed contractual interest rate can be used to gain a higher rate of return than the interest rate paid. The concept of financial leverage is borrowing to increase the return on equity by maximizing the difference between the rate of return and the interest paid.

Now let's take a look at financial leverage in more detail. First consider the return on equity, R, in relation to profit after taxes, Z, and equity, E, as shown in Equation (9.4).

$$R = \frac{Z}{E} \tag{9.4}$$

The return on equity is simply the profit after taxes divided by equity. Next consider capitalization. It is the sum of equity and debt, D.

$$\text{Capitalization} = E + D \tag{9.5}$$

Profit after taxes can now be considered as shown in Equation (9.6).

$$Z = r(E + D) - Di \tag{9.6}$$

In Equation (9.6), r is the return on capitalization, and i is the interest rate paid on the debt. The return on capitalization is the return achieved by the operations area using the money received from the investment area. The profit is the difference between the return on capitalization and the cost of interest on the outstanding debt. The return on equity can be restated by substituting Equation (9.6) into Equation (9.4) and rearranging.

$$R = r + \frac{D}{E}(r - i) \tag{9.7}$$

Equation (9.7) is the key formula to describe financial leverage. The leverage effect is represented by the ratio of debt to equity multiplied by the difference between the rate of return on capitalization and the rate of interest paid. Thus, as the debt to equity ratio increases, the return on equity is enhanced as long as the earnings power exceeds interest costs. Of course the converse is also true. As the debt to equity ratio increases, losses on equity increase when interest costs exceed the earnings power.

Figure 9.3 illustrates the effect of financial leverage by utilizing Equation (9.7). In this example, the interest on the debt is assumed to be 6%. Figure 9.3 shows how return on equity is a function of return on capitalization and the debt to equity ratio. Note that as increasing amounts of debt are introduced to the capital structure, the return on

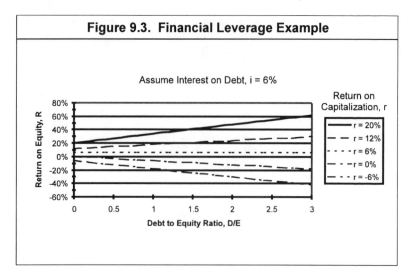

Figure 9.3. Financial Leverage Example

equity is boosted either positively or negatively depending on whether or not the return on capitalization exceeds the interest cost.

Chapter 10

FINANCIAL REPORTING

The last chapter presented a simple model showing how different areas of the business fit together. There is the operations side which balances price, cost, and volume to achieve an operating profit. Then there is the investment side which dispositions some of the operating profit to owners and lenders. The remaining profit is balanced with long-term debt in order to make new investments back into the operations.

This chapter covers how the financial score is kept, reporting how the business is doing in the operations and investment areas. The primary financial reports to understand are the income statement and the balance sheet.

The income statement shows how well the business did this year compared to last year. It tells how much of a profit or loss was generated. The balance sheet shows the strength of the business' finances by stating what it owns and owes on a certain date. Table 10.1 summarizes the definitions of financial reporting terms.

To aid in this explanation, the finances of an actual company will be used. This company is Quantum Corporation. Quantum Corporation designs, manufactures, and markets high capacity, hard disk drives for the computer industry. The computer industry was hit with price wars for its products in the early 1990's, and this led to decreasing prices in the disk drive industry. It was imperative for disk drive companies to introduce new products and to trim expenses in order to maintain profit margins.

Table 10.1. Glossary of Financial Reporting Terms

Accounts payable: Debts owed to other companies for the purchase of goods or services.

Accounts receivable: Money yet to be collected from customers.

Assets: All items owned including claims due from others.

Balance sheet: A financial statement which shows what the business owns and owes on a certain date.

Bond: A certificate showing evidence of a loan with interest usually paid twice a year.

Break even point: The volume of business where profits equal zero. At this point, revenues offset the fixed and variable costs.

Capital stock: Shares of the company consisting of stock certificates issued to stockholders. These shares can be preferred stock or common stock.

Capital surplus: The amount paid by stockholders over the par value of each share when the shares were issued.

Common stock: Shares of the company with lower priority than preferred stock regarding receipt of dividend payments.

Cost of goods sold: See "Cost of sales".

Cost of sales: The expenses incurred to purchase raw materials and convert them into finished goods.

Current assets: Cash and those assets expected to be turned into cash in the near future, usually within one year of the balance sheet date.

Current liabilities: Debts which become due within the next year.

Debt: Money owed to others.

Depreciation: The decline in value of a fixed asset over time due to either wearing out or otherwise losing its usefulness to the business.

Equity: See "Stockholders' equity".

Financial leverage: Profits rise proportionately faster when borrowed funds are used to earn more money than the interest payments for the borrowed funds. Also losses mount proportionately faster if the rate of return from the borrowed funds is less than the interest paid.

Fixed assets: Items owned and used repeatedly over several years to manufacture and sell products. Examples include land, buildings, and machinery.

Fixed costs: Costs which do not change as the volume of product made varies.

Gross profit: The difference of "net sales" minus "cost of sales".

Table 10.1. (continued)

Income statement: A statement showing how much the business made or lost during the statement period.

Income tax provision: The amount due to the government on the taxable income earned during the income statement period.

Inventory: Items stored such as raw materials used to make the product, partially finished goods in the process of being manufactured, and finished goods ready to be sold.

Liabilities: All debts owed to others.

Long-term liabilities: Debts due more than one year after the balance sheet date.

Net income: See "Net profit".

Net profit: Calculated by considering all income and deducting all costs and expenses for the income statement period. The difference of "profit before taxes" minus the "income tax provision".

Net sales: The revenue received for goods sold or services rendered after considering sales returns, discounts off list prices, and prompt payment discounts.

Operating expenses: Costs incurred relating to research and development, sales and marketing, and administration.

Operating leverage: Profits rise proportionately faster than the increase in business volume above the break even point, and losses mount proportionately faster than the decrease in business volume below the break even point.

Operating profit: The difference of "gross profit" minus "operating expenses".

Preferred stock: Shares of the company with preference over common stock with regards to dividend payments and in the event of liquidation, distribution of assets.

Profit before taxes: The amount obtained from "operating profit" plus "interest and other income" minus "interest expense".

Retained earnings: The sum of this year's profits, after payment of dividends, added to the profits accumulated since the beginning of the business.

Sales: The revenue received from customers for goods sold or services rendered.

Stockholders' equity: The amount owners would get if the business was liquidated at the values shown on the balance sheet. This is the net worth of the business.

Variable costs: Costs which are a function of the volume produced.

1. BALANCE SHEET

The balance sheet represents the financial condition of the business on a certain date. In the case of Quantum, the most recent balance sheet date shown in Table 10.2 is March 31, 1993. The balance sheet includes not only the most recent fiscal year, but also the previous year so comparisons can be made. All dollar figures in the financial tables are expressed in thousands of dollars to make the numbers easier to read.

The most fundamental relationship of a balance sheet is that the assets are always balanced with the liabilities and stockholders' equity.

$$\text{Assets} = \text{Liabilities} + \text{Stockholders' Equity} \qquad (10.1)$$

Assets: all items owned including claims due from others

Liabilities: all debts owed to others

Stockholders' equity: the amount owners would get if the business was liquidated at the values shown on the balance sheet

Assume the firm goes out of business on the balance sheet date. Also assume the assets are sold at values shown on the balance sheet. For Quantum, the stockholders would expect to receive the following.

Total assets	$926,633
Total liabilities	528,395
Amount remaining for the stockholders	$398,238

Next is an explanation of each of the balance sheet entries.

1.1. Assets

1.1.1. Current Assets

Current assets consist of cash and those assets expected to be turned into cash in the near future, usually within one year of the balance sheet date.

Cash

There is no surprise here. This is money deposited in the bank and money on hand in the petty cash fund. Cash equivalents are highly liquid investments with maturities of three months or less from the date of acquisition.

Table 10.2. Balance Sheet for Quantum Corporation[27]

(thousands of dollars)

	March 31, 1993	March 31, 1992
Assets		
Current assets		
Cash and cash equivalents	$121,838	$74,486
Marketable securities	$167,114	$70,547
Accounts receivable, net of allowance for doubtful accounts of $8,118 and $6,474	$266,994	$194,350
Inventories	$223,162	$87,375
Deferred tax assets	$37,479	$21,958
Other current assets	$13,094	$5,919
Total current assets	$829,681	$454,635
Fixed assets		
Property and equipment, at cost less accumulated depreciation	$74,698	$65,831
Net fixed assets	$74,698	$65,831
Other assets (including investments)	$22,254	$30,398
Total assets	$926,633	$550,864
Liabilities		
Current liabilities		
Accounts payable	$215,445	$171,346
Accrued warranty expense	$42,410	$29,567
Accrued compensation	$17,189	$9,583
Income taxes payable	$19,026	$16,898
Other accrued liabilities	$21,825	$15,081
Total current liabilities	$315,895	$242,475
Long-term liabilities		
Subordinated debentures	$212,500	$0
Commitments and contingencies	$0	$0
Total long-term liabilities	$212,500	$0
Total liabilities	$528,395	$242,475
Stockholders' equity		
Preferred stock, $.01 par value, 4,000,000 shares authorized, none issued	$0	$0
Common stock, $.01 par value, 150,000,000 shares authorized, 43,321,588 and 42,893,007 shares outstanding	$433	$429
Capital in excess of par value	$99,616	$88,582
Retained earnings	$298,189	$219,378
Total stockholders' equity	$398,238	$308,389
Total liabilities and stockholders' equity	$926,633	$550,864

---

27.   Quantum Corporation 1993 Annual Report and Form 10-K.

Cash and cash equivalents $121,838

Marketable securities

These are investments of cash not needed right away. A common investment is in short-term government securities. These securities may be needed quickly, so they must be easily sellable and not fluctuate much in price. In most cases, these are shown on the balance sheet at fair market value.

Marketable securities $167,114

Accounts receivable

This reflects the money yet to be collected from customers. These are customers who received goods before paying. They are usually given 30, 60, or 90 days to pay their bills. Also reflected in the balance sheet is the projected deduction in accounts receivable for some customers who fail to pay due to financial difficulties.

Accounts receivable, net of allowance for $266,994
   doubtful accounts of $8,118

Inventories

For a company such as Quantum that manufactures products, inventory is composed of three groups:

- raw materials used to make the product

- partially finished goods in the process of being manufactured

- finished goods ready to be sold

To be conservative, the usual method of valuation is cost or market value, whichever is lower. Thus if the products in inventory are obsolete or have declining market prices, their valuation for balance sheet purposes can be less than the manufactured cost.

Inventories $223,162

Other current assets

Other current assets can include prepaid expenses and deferred charges. Examples of prepaid expenses are insurance premiums and advertising expenses. These expenses are prepaid and may be unused at the time of the balance sheet date. However they are expected to be used over the next 12 months. If the company had not made advance payments, there would have been more money in the bank. Thus they are included as current assets.

Deferred charges are assets similar to prepaid expenses. However, the benefits of deferred charges may be spread over periods longer than one year. Thus only those deferred charges that are expected to be used up within a year are counted as current assets, and the rest is written off gradually over the next several years. Examples of deferred charges are taxes, introducing a new product to market, and moving a plant to a new site.

Deferred tax assets	$37,479
Other current assets	$13,094

In summary, current assets mainly consist of cash, marketable securities, accounts receivable, and inventories. They are continually being converted into cash, and this cash is used partially to pay debts and expenses.

Total current assets	$829,681

1.1.2. Fixed Assets

Fixed assets are long-term operating assets used by the business over several years. They are used repeatedly to manufacture and sell the product. Fixed assets can include land, buildings, machinery, furniture, and trucks.

The valuation used for fixed assets is generally cost minus the depreciation accumulated by the balance sheet date. Depreciation is defined as the decline in value of a fixed asset due to either wearing out or otherwise losing its usefulness to the business. For example, suppose a computer system is purchased for $20,000 and is expected to last five years. Using the straight line method of depreciation, the computer system will be depreciated by $4000 each year. The balance sheet after one year would show:

Computer system (cost)	$20,000
Less accumulated depreciation	4,000
Net depreciated value	$16,000

After the second year, the balance sheet would show:

Computer system (cost)	$20,000
Less accumulated depreciation	8,000
Net depreciated value	$12,000

As a special note, land is not subject to depreciation, and its value listed on the balance sheet remains unchanged from year to year.

The valuation of fixed assets is not intended to represent current market value or projected replacement costs. The replacement cost will probably be higher, but this is difficult to judge. Thus, it is simpler to report cost minus accumulated depreciation on the balance sheet.

Property and equipment, at cost less accumulated depreciation	$74,698

1.1.3. Other Assets

Other assets can include investments. Also included can be intangibles that may have substantial value. Examples of intangibles are patents and well known trademarks.

Other assets (including investments)	$22,254

1.1.4. Total Assets

Total assets are the sum of the current, fixed, and other assets.

Total Assets	$926,633

1.2. Liabilities

1.2.1. Current Liabilities

Current liabilities are those debts which become due within the next year. They have a close relationship with current assets. Remember current assets are those assets which are cash or expected to be converted into cash in the next year. This cash is the source of payment for the current liabilities.

Accounts payable

Accounts payable are the payments the business owes other companies for the purchase of goods or services.

Accounts payable $215,455

Accrued expenses payable

Besides accounts payable owed to other companies, the business also owes at any given time salaries and wages to its employees, interest on borrowed funds from banks, insurance premiums, and other items. Amounts owed that are unpaid on the balance sheet date are listed in this category.

Accrued compensation $17,189
Accrued warranty expense $42,410

Income tax payable

Amounts owed for income taxes are the same as other accrued expenses payable. However they are listed separately because of the importance of being able to quantify tax effects.

Income taxes payable $19,026

Other accrued liabilities

Other accrued liabilities $21,825

Total current liabilities consist of accounts payable, accrued expenses payable, income taxes payable, and other accrued liabilities.

Total current liabilities $315,895

1.2.2. Long-Term Liabilities

Long-term liabilities are those debts due more than one year after the balance sheet date.

Debentures

A common long-term liability is the money owed to bondholders. Companies receive money as a loan from bondholders. The bondholders are given a bond which is a certificate showing evidence of the loan. Interest on the bonds is usually paid twice a year.

Bond issues are called debentures when they are backed by the general credit of the company rather than its assets. When well-established corporations issue bonds, debentures are the most common type. Another type of bond is a first mortgage bond. This type of bond is backed by a mortgage on all of the company's property and is considered one of the highest grade investments.

In the case of Quantum, they issued 6 3/8% convertible subordinated debentures in April 1992. These bonds pay an interest at a rate of 6 3/8% per year. They are called convertible because there is a provision which allows bondholders to convert their bond into common stock at a specified price if they so choose.

Subordinated debentures	$212,500

Commitments and contingencies

Commitments and contingencies are always reported as zero on the balance sheet. Its purpose is to alert readers of potential liabilities that are not recorded as of the balance sheet date. Any commitments and contingencies are described in the footnotes to the balance sheet. Examples are purchase commitments, unused lines of credit, and contingent liabilities such as litigation claims or environmental clean-up costs. Such commitments and contingencies are considered to be reasonably possible, but the amounts are not determinable due to uncertainty of the outcome.

Commitments and contingencies	$0

1.2.3. Total Liabilities

Total liabilities are the sum of the current and long-term liabilities.

Total Liabilities	$528,395

1.3. Stockholders' Equity

This is the total equity stockholders have in the corporation. It is the corporation's net worth after subtracting out all the liabilities.

1.3.1. Capital Stock

This represents shares of the company, and these shares consist of stock certificates which the corporation issues to its stockholders. There are different types of shares.

Preferred stock

These shares have preference over other types of capital stock with regards to dividend payments and distribution of assets in the event of liquidation. Quantum's Board of Directors authorized 4,000,000 shares of preferred stock but none had been issued as of the balance sheet date.

Preferred stock, $0.01 par value, 4,000,000 shares authorized, none issued	$0

Common stock

When the company pays out dividends, common stockholders have lower priority than preferred stockholders. The preferred stockholders are paid their dividends first. Then if the company deems it appropriate to pay out additional dividends, the common stockholders receive theirs. Dividends to common stockholders can fluctuate depending on the earnings of the corporation. And in many cases such as Quantum, where growth is a primary objective, no dividends are paid to common stockholders. With no dividend payments, more funds are available for new investment back into the company's operations to fuel more growth.

Common stock, $0.01 par value, 150,000,000 shares authorized, 43,321,588 shares outstanding	$433

1.3.2. Capital Surplus

Capital surplus is the amount paid by stockholders over the par value of each share when the shares were issued.

Capital in excess of par value	$99,616

1.3.3. Retained Earnings

This is also called accumulated retained earnings. The balance sheet reflects the profits for the corporation after dividends have been paid, and this is added to the previous year's balance. When a corporation first begins, there are no accumulated retained earnings. Hypothetically, if net profits are $10,000 for the first year and $4,000 are paid in dividends, the accumulated retained earnings are $6,000 at the end of the first year. If in the second year, the net profits are $15,000 and dividends paid out are $5,000, the accumulated retained earnings at the end of the second year are $16,000.

| Retained earnings | $298,189 |

1.3.4. Total Stockholders' Equity

This is the sum of the capital stock, capital surplus, and retained earnings.

| Total Stockholders' Equity | $398,238 |

And note how the balance sheet is balanced. As pointed out in Equation (10.1), the total assets equals the sum of the total liabilities and total stockholders' equity.

2. INCOME STATEMENT

The income statement shows how much the corporation either made or lost during the year. While the balance sheet shows the financial condition of the business on a given date, engineers may relate easier to the income statement. This is because the income statement shows how well the operations side of the business performed over the year. An income statement summarizes the revenue received from selling goods and services and summarizes the expenses due to the operation of the business. The result is a net profit or a net loss for the year. Quantum's income statement for the 1992 fiscal year is shown in Table 10.3.

2.1. Net Sales

The first item on the income statement always represents the most important source of revenue received from customers for goods sold or services rendered. Net sales is the amount received after considering sales returns, discounts off list prices, and prompt payment discounts.

| Net sales | $1,697,240 |

2.2. Cost of Sales

This is also known as cost of goods sold. It represents the costs incurred to purchase raw materials and convert them into finished goods. These costs include raw material costs, direct labor costs, and manufacturing overhead. Direct labor costs are the labor costs of the

Table 10.3. Income State for Quantum Corporation[28]

(thousands of dollars)

	1992*	1991*
Net sales	$1,697,240	$1,127,733
Cost of sales	$1,374,422	$914,348
Gross profit	$322,818	$213,385
Operating expenses		
Research and development	$63,019	$59,255
Sales and marketing	$77,085	$55,027
General administrative	$33,849	$23,852
Total operating expenses	$173,953	$138,134
Operating profit	$148,865	$75,251
Interest and other income	$12,077	$6,868
Interest expense	($14,363)	($7,763)
Profit before income taxes	$146,579	$74,356
Income tax provision	$52,768	$27,511
Net profit	$93,811	$46,845

* 1992 is the fiscal year which ended 3/31/93.
  1991 is the fiscal year which ended 3/31/92.

people whose work can be directly attributable to producing the product, such as machine operators and assemblers. Manufacturing overhead represents those costs associated with operating the manufacturing facility such as supervision, rent, electricity, maintenance, and depreciation. As mentioned earlier, depreciation measures the decline in value of a fixed asset due to wear and tear. Each year's decline in value of a machine used in the manufacturing process is a cost to be included in the manufacturing overhead.

Cost of sales $1,374,422

2.3. Gross Profit

The difference between net sales and cost of sales is called the gross profit.

28.   Quantum Corporation 1993 Annual Report and Form 10-K.

Net sales	$1,697,240
Cost of sales	1,374,422
Gross profit	$322,818

2.4. Operating Expenses

These expenses are generally listed separately from cost of sales so that the extent of development, selling, and administrative costs can be identified. Included here are salaries and other expenses of those engineers involved in research and development, salaries and commissions of sales personnel, executives' salaries, office payroll, and expenses for advertising, travel, and office equipment.

Operating expenses	
Research and development	$63,019
Sales and marketing	77,085
General and administrative	33,849
	$173,953

2.5. Operating Profit

Subtracting operating expenses from gross profit gives the operating profit. This is the profit generated during the year by the operations area.

Gross profit	$322,818
Operating expenses	173,953
Operating profit	$148,865

2.6. Interest and Other Income

The rest of the income statement involves the investment area of the business rather than the operations area. Another source of revenue comes from the company's investments in stocks and bonds. These are in the form of dividends and interest received by the company.

Interest and other income	$12,077

2.7. Interest Expense

Interest paid to bondholders and other creditors for use of their money is another cost of doing business. Interest differs from dividends paid to the stockholders. Interest expenses are deductible from operating profits in order to calculate the taxable earnings for income tax purposes. Dividends paid by the corporation are not deductible from its taxable income.

Interest expense	$14,363

2.8. Profit Before Taxes

This is obtained by adding interest and other income to the operating profit and subtracting interest expense.

Operating profit	$148,865
Interest and other income	12,077
Interest expense	(14,363)
Profit before taxes	$146,579

2.9. Income Tax Provision

This is the amount due to the government on the taxable income earned by the company during the year. It is calculated by multiplying the taxable income by the appropriate tax rate. The tax rate depends on the level of income. The income tax provision can sometimes be lowered by the use of tax credits and depreciation write-offs.

Income tax provision	$52,768

2.10. Net Profit

This is also known as net income. It is calculated by considering all income and deducting all costs and expenses for the year. Following the flow of the income statement, net profit is found by subtracting the income tax provision from the profit before taxes.

Profit before taxes	$146,579
Income tax provision	52,768
Net profit	$93,811

Chapter 11

FINANCIAL ANALYSIS

Chapter 9 presented a model showing how various areas of the business fit together, and Chapter 10 explained financial reporting. Chapter 9 briefly mentioned key financial measures without much explanation. Now that we have a better understanding of financial reporting, we can now explore the key measures in greater detail. Part of the presentation in this chapter will be an explanation of financial linkage analysis. This tool presents an overall view of how business finances are tied together. Financial linkage analysis will serve as the basis for you to understand how your activities can impact the key financial measures. Table 11.1 summarizes the financial measures and equations that will be presented in this chapter.

1. FINANCIAL HEALTH

This section presents some of the key measures from the business model related to financial health. These are certainly not the only measures of financial health, but it provides a good starting point for financial analysis.

1.1. Net Working Capital

One important item to be learned from the balance sheet is net working capital, also known as net current assets or working capital.

Net working capital = Current assets - Current liabilities   (11.1)

Recall current liabilities are those liabilities due within one year of the balance sheet date. The source of funds used to pay these debts is the current assets. Therefore net working capital corresponds to the

Table 11.1. Summary of Financial Measures

I. Financial Health

Net working capital = Current assets - Current liabilities

$$\text{Current ratio} = \frac{\text{Current assets}}{\text{Current liabilities}}$$

$$\text{Interest coverage} = \frac{\text{Operating profit} + \text{Other income}}{\text{Interest expense}}$$

II. Financial Linkage

$$\text{Gross profit margin (\%)} = \frac{\text{Sales} - \text{Cost of sales}}{\text{Net sales}} = \frac{\text{Gross profit}}{\text{Net sales}}$$

$$\text{Operating profit margin (\%)} = \frac{\text{Gross profit} - \text{Operating expenses}}{\text{Net sales}} = \frac{\text{Operating profit}}{\text{Net sales}}$$

$$\text{Profit margin, before taxes (\%)} = \frac{(\text{Operating profit} + \text{Other income}) - (\text{Interest expense})}{\text{Net sales}}$$

$$= \frac{\text{Profit before taxes}}{\text{Net sales}}$$

Total assets = Current assets + Fixed assets + Other assets

$$\text{Asset turnover} = \frac{\text{Net sales}}{\text{Total assets}}$$

Return on assets, before taxes (%) = (Profit margin, before taxes) × (Asset turnover)

Total liabilities = Current liabilities + Long term liabilities

Stockholders' equity = Capital stock + Capital surplus + Retained earnings

Capitalization = Total liabilities + Stockholders' equity

Table 11.1. Summary of Financial Measures (continued)

II. Financial Linkage (continued)

$$\text{Leverage factor} = \frac{\text{Total liabilities} + \text{Stockholders' equity}}{\text{Stockholders' equity}} = \frac{\text{Capitalization}}{\text{Stockholders' Equity}}$$

Return on equity, before taxes (%) = (Return on assets, before taxes) × (Leverage factor)

Tax factor (%) = 1 - Tax rate

Return on equity, after taxes (%) = (Return on equity, before taxes) × (Tax factor)

$$\text{Earnings retention rate (\%)} = 1 - \frac{\text{Dividends}}{\text{Net profits, after taxes}}$$

Sustainable rate of growth (%) = (Return on equity, after taxes) × (Earnings retention rate)

$$= \frac{\text{Net profits, after taxes} - \text{Dividends}}{\text{Stockholders' equity}}$$

---

amount left after all current liabilities are paid off. In the case of Quantum, this was:

Current assets	$829,681
Less: Current liabilities	315,895
Net working capital	$513,786

	3/31/93	3/31/92
Net working capital	$513,786	$212,160

Both years presented in the balance sheet should be examined to see if working capital is growing or shrinking. Healthy companies keep a safe cushion of working capital. The company's ability to pay off debts, grow the business, and take advantage of new opportunities is often judged by the amount of its working capital. If one of the objectives for the company is to grow, it is desirable for this year's working capital to be larger than last year's. In the case of Quantum, its working capital more than doubled. This was a sign of a company growing quickly.

1.2. Current Ratio

This ratio helps to define a comfortable amount of working capital. The most common method to help interpret the current financial position of an industrial company is to calculate the current ratio. This is a key measure of the operations area from the business model.

$$\text{Current ratio} = \frac{\text{Current assets}}{\text{Current liabilities}} \qquad (11.2)$$

There are many exceptions, but generally, minimum safety requires the current ratio to be at least 2. A current ratio of 2 means the value of current assets is twice the value of current liabilities. A current ratio of 2 means the company can pay off its current liabilities and still have some current assets left. Companies that have little inventory and can easily collect on their accounts receivable may get by with a lower current ratio than companies that have a larger portion of current assets in inventory and sell more products on credit.

For Quantum, the current ratio was:

$$\text{Current ratio} = \frac{\$829,681}{\$315,895} = 2.63$$

	3/31/93	3/31/92
Current ratio	2.63	1.88

As of 3/31/93, the current ratio was 2.63, meaning for each $1 of current liabilities, there was $2.63 in current assets to back it up. Quantum's current ratio strengthened considerably from a year earlier. Thus when using working capital and current ratio as indicators, Quantum was not only growing, but it was becoming stronger financially.

1.3. Interest Coverage

Interest coverage is another key measure from the business model. This measure comes from the investment area of the business. Recall in the business model that operating profits can be dispositioned by paying interest to cover debt, paying dividends to stockholders, or by retaining earnings.

The bonds and other borrowings of a corporation can represent a very substantial debt, and the annual interest on this debt is a fixed charge. It is desirable to know if these borrowed funds were used to generate enough earnings to meet the interest costs. The interest coverage is the ratio between available income and the interest expense. The available income is the income from operations plus interest and other income.

$$\text{Interest coverage} = \frac{\text{Operating profit} + \text{Other income}}{\text{Interest expense}} \quad (11.3)$$

$$\text{Interest coverage} = \frac{\$148,865 + \$12,077}{\$14,363} = 11.2$$

	3/31/93	3/31/92
Interest coverage	11.2	10.6

The interest coverage for Quantum shows the annual interest expense was covered 11.2 times. Of note is that during the latest fiscal year, Quantum issued new bonds which increased its interest expense. However the interest coverage actually increased from 10.6 to 11.2 which indicated that the money borrowed by issuing the bonds was put to good use.

Before a bond is viewed as a safe investment, many financial analysts require the company's interest coverage to be at least three to four. By these standards, Quantum's interest coverage showed a good margin of safety.

2. FINANCIAL LINKAGE

Financial linkage analysis is a method to show how key measures are linked together and how key measures can be affected in order to create a sustainable rate of growth. The analysis links the operations and investment areas as well as information from the balance sheet and income statement. This tool shows how your activities can change the key financial measures of the business, ultimately leading to the sustainable rate of growth.

Figure 11.1 shows a blank form for financial linkage analysis. This figure can be copied so you can analyze your own company and your competitors. Table 11.1 summarizes how each calculation in this analysis is made. As an example, Figure 11.2 shows this analysis for Quantum Corporation. Figure 11.2 labels Quantum's fiscal years which ended on 3/31/93 and 3/31/92 as 1992 and 1991, respectively.

2.1. Profit Margin

Profit margin is a key measure for the operations area. The most direct measure of the operations is the operating profit margin, which is simply the operating profit divided by net sales. The operating profit margin can be analyzed further by realizing it is tied to gross profit margin and operating expenses.

$$\text{Operating profit margin (\%)} = \frac{\text{Gross profit - Operating expenses}}{\text{Net sales}} \quad (11.4)$$

For Quantum, the operating profit margin improved from 6.7% in 1991 to 8.8% in 1992. Further analysis shows that this improvement was not due to gross profit margin improvement because the gross profit margin was almost the same. The improvement in operating profit margin came directly from a reduction of operating expenses as a percent of net sales from 12.2% in 1991 to 10.2% in 1992.

Another profit margin used in financial linkage analysis is profit margin before taxes. This is the profit before taxes divided by net sales. Profit before taxes is calculated starting with the operating profit, adding other income, and subtracting interest expenses.

$$\text{Profit margin, before taxes (\%)} = \frac{(\text{Operating profit + Other income}) - (\text{Interest expense})}{\text{Net sales}} \quad (11.5)$$

In Quantum's case, the profit margin before taxes increased from 6.6% in 1991 to 8.6% in 1992. Again this improvement could be traced to the decrease in operating expenses as a percent of net sales.

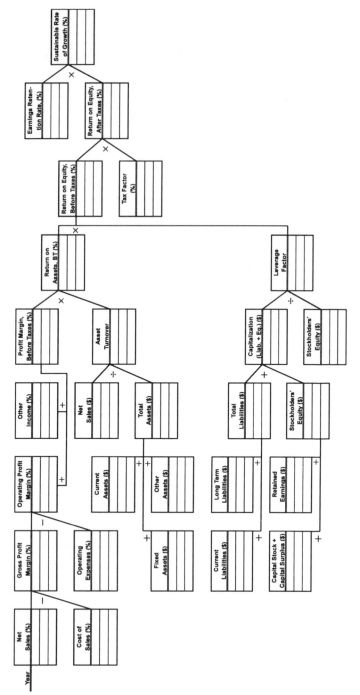

Figure 11.1. Form for Financial Linkage Analysis

Figure 11.2. Financial Linkage Analysis for Quantum Corporation[29,30]

All dollar figures in thousands of dollars

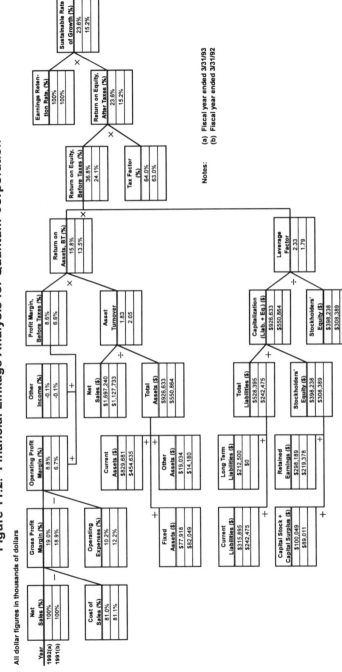

Notes: (a) Fiscal year ended 3/31/93
         (b) Fiscal year ended 3/31/92

29. Quantum Corporation 1993 Annual Report and Form 10-K.
30. George A. Niemond, "Quantum Corporation," The Value Line Investment Survey, Apr. 30, 1993, 1103.

2.2. Return on Assets

This is another key measure of the operations. Return on assets is also known as return on operations. This measures how effectively assets are utilized to achieve profits. It is defined as profit margin before taxes multiplied by asset turnover. Asset turnover is net sales divided by total assets and can be viewed as an efficiency rating for utilizing assets to achieve sales. Higher profit margins and higher asset turnover increase the return on assets.

Return on assets, before taxes (%) = (Profit margin, before taxes) × (Asset turnover) (11.6)

For Quantum, return on assets before taxes increased from 13.5% to 15.8%. This increase was due to a better profit margin before taxes (6.6% in 1991 and 8.6% in 1992). However the improvement in profit margin was partially offset by a lower asset turnover. Let's take a closer look at the components for asset turnover. Going from 1991 to 1992, Quantum's sales increased substantially from $1.13 billion to $1.70 billion. Its total assets also increased because current assets increased from $455 million to $830 million. When there is such a large growth in assets, it is sometimes difficult to maintain the same level of asset utilization efficiency. The balance sheet shows the main reason for current assets increasing was due to increased inventories. Inventories rose from $87 million in 1991 to $223 million in 1992. In order to increase asset turnover to improve return on assets, one area to focus on would be to reduce inventory levels relative to net sales.

2.3. Capitalization to Equity Ratio (Financial Leverage)

With financial leverage, there is an advantage when funds borrowed at a fixed interest rate can be used to gain a higher rate of return than the interest paid. But there is a disadvantage when the rate of return is lower than the interest paid. Thus with higher financial leverage, there is opportunity for greater returns, but the risks are also greater. Key measures of financial leverage are the ratio of debt to equity and the ratio of total capitalization (debt plus equity) to equity. For the purposes of financial linkage analysis, we will focus on the latter ratio. The latter ratio is called the leverage factor.

$$\text{Leverage factor} = \frac{\text{Total liabilities} + \text{Stockholders' equity}}{\text{Stockholders' equity}} \quad (11.7)$$

For Quantum, the leverage factor increased from 1.79 to 2.33. By analyzing the financial linkage, it can be seen the main reason the leverage factor increased was because long-term liabilities went from zero to $212.5 million. Recall that in 1992, Quantum issued bonds at 6 3/8%. Quantum issued these bonds in order to increase its financial leverage. Next, we will see how this impacted return on equity.

2.4. Return on Equity

Financial linkage shows clearly how return on equity is calculated and how financial leverage affects it. The return on equity before taxes is the product of return on assets before taxes and the leverage factor. To find the return on equity after taxes, just multiply the return on equity before taxes by the tax factor.

$$\text{Return on equity, before taxes (\%)} = (\text{Return on assets, before taxes}) \times (\text{Leverage factor}) \quad (11.8)$$

$$\text{Return on equity, after taxes (\%)} = (\text{Return on equity, before taxes}) \times (\text{Tax factor}) \quad (11.9)$$

Equation (9.4) showed that the same result for return on equity after taxes can be obtained by calculating profit after taxes divided by equity.

Quantum's return on equity increased significantly in 1992 compared to 1991. Looking at return on equity before taxes, it increased from 24.1% to 36.8%. This shows the compounding effect, on the positive side, of increased financial leverage. Quantum's issuance of bonds at 6 3/8% increased the amount of assets for the operations area. The operations achieved a higher return with this money than the interest paid on the bonds. As shown earlier, return on assets before taxes increased from 13.5% to 15.8% in 1992. The higher return on assets multiplied by a higher leverage factor led to significantly higher return on equity.

2.5. Earnings Retention Rate

This is the fraction of net profits after taxes that remains after dividends have been paid.

$$\text{Earnings retention rate (\%)} = 1 - \frac{\text{Dividends}}{\text{Net profits, after taxes}} \quad (11.10)$$

Typically for companies where high growth is a primary objective, no dividends are paid. This maximizes the amount of earnings that can be invested back into the business as shown in the business model. Quantum paid no dividends, so its earnings retention rate was 100%.

2.6. Sustainable Rate of Growth

Sustainable rate of growth is the "bottom line" indicator for financial linkage analysis. All components in this analysis tie into calculating sustainable rate of growth. Sustainable rate of growth is the rate at which equity grew for the current year, and it is the rate at which the business will grow at the current earnings retention rate assuming the same return on equity.

$$\text{Sustainable rate of growth (\%)} = \frac{\text{Net profits, after taxes - Dividends}}{\text{Stockholders' equity}} \quad (11.11)$$

For Quantum, its sustainable rate of growth was the same as the return on equity after taxes because its earnings retention rate was 100%. Quantum's sustainable rate of growth increased significantly from 15.2% to 23.6%. The beauty of financial linkage analysis is that the causes of this change in sustainable rate of growth can be seen easily. The reasons for improved financial performance were mainly the decrease in operating expenses relative to sales and taking on long-term liabilities. The decrease in operating expenses relative to sales led to a higher operating profit margin which led to a higher profit margin which led to a higher return on assets. The issuance of bonds led to a higher financial leverage factor. Quantum was able to take advantage of the increased financial leverage.

Financial linkage analysis also points out areas for possible improvement. For Quantum, one area for possible improvement was in gross profit margin. Quantum was in a very difficult market because of rapid technical obsolescence and declining prices for its products. The disk drive market was in the midst of a price war.[31] Thus gross margins continued to come under pressure. Quantum's strategy focused on introducing quickly higher capacity disk drives which had better gross margins than lower capacity disk drives.[32] Engineers impacted

---

31. George A. Niemond, "Computers and Peripherals Industry," *The Value Line Investment Survey,* Apr. 30, 1993, 1076.

the finances most significantly by developing such new products in a timely manner. Another area of possible improvement mentioned earlier was to decrease inventory relative to sales. This would help improve asset turnover which would improve return on assets. However this again points out a difficulty that faced the disk drive industry. Reducing inventories usually meant fire sale prices on older, lower capacity disk drive products. This put even more pressure on gross margins.

Note how this coverage of financial linkage analysis did not go directly to sales and profits and quit there. In the case of Quantum, it was enormously successful in 1992 compared to 1991. Its sales grew from $1.13 billion to $1.70 billion, and its net profits after taxes doubled from $46.8 million to $93.8 million. However, it is important for engineers to understand how the components of the financial system fit together in order to impact more directly critical areas of the business. Sales and net profits are important, but it is also important to recognize rate of growth and how rates of growth can be sustained for long-term business success.

The most effective methods of financial analysis are to make year by year comparisons of the business as well as making comparisons with companies competing in the same industry. The next two chapters present case studies, so you can see more examples of financial analysis.

32. George A. Niemond, "Quantum Corporation," *The Value Line Investment Survey,* Apr. 30, 1993, 1103.

Chapter 12

CASE STUDY—QUANTUM AND MICROPOLIS

Quantum and Micropolis are two companies that compete in the computer disk drive market. This business is very volatile because of rapid advances in technology. In general, orders for computer equipment increased significantly in 1992. However, the problem for computer manufacturers was that price wars had broken out in the industry for all computers, from personal computers to mainframes. This price pressure spread to suppliers such as disk drive manufacturers. The rapid decline in prices meant more units had to be shipped to keep sales (in terms of dollars) the same. Even more units needed to be sold to increase sales. The price declines also drove profit margins lower. In this environment, engineers faced the increasingly difficult challenge of differentiating their products on the basis of performance, reliability, and service. Engineers affected company finances most significantly by introducing new products with higher margins more rapidly.

Micropolis develops and manufactures high performance disk drives, and markets them to original equipment manufacturers (OEMs).[33] In 1992, Micropolis' principal products were 3½ inch and 5¼ inch Winchester disk drives with high capacity and fast access times. These were used in high performance microcomputers and minicomputers such as engineering workstations and local area network file servers.

---

33. Lucien Virgile, "Micropolis Corporation," *The Value Line Investment Survey,* Apr. 30, 1993, 1101.

The disk drive business for microcomputers and minicomputers was undergoing dramatic change. The demand for 3½ inch high performance Winchester disk drives was rapidly overtaking demand for 5¼ inch drives used in workstations and file servers. It was estimated that drives accounting for well over half of Micropolis' revenues in 1992 would no longer be sold in 1993 and beyond. Almost all of the company's 5¼ inch drives with capacities less than 1.6 GBytes were to be replaced by 3½ inch models. Sales in 1993 were to come mainly from products introduced in 1991 and later. This was quite an example of technological change!

Quantum competed in a similar but slightly different disk drive market. It also developed and manufactured disk drives based on Winchester technology, and marketed them to OEMs. However, Quantum's principal products were 3½ inch and 2½ inch disk drives, and these were used mainly in personal and notebook computers.

Like Micropolis, Quantum faced intense pricing pressure and rapid technological change for its products. Users were rapidly moving to higher capacity disk drives as larger operating systems and applications used up more disk space. Quantum's strategy was focused on bringing to market as quickly as possible new products with higher capacities and higher profit margins while offering fire sale prices on its older, lower capacity drives.

1. FINANCIAL SUMMARY AND FINANCIAL HEALTH ANALYSIS

Table 12.1 shows the pertinent information for this discussion. Quantum was quite successful in 1992 compared to 1991. Its sales increased 50%, and its net profits doubled. The sustainable rate of growth increased significantly from 15.2% to 23.6%.

In terms of financial health, Quantum was able to improve its position in 1992. Recall that for a growing company, it is desirable to have growing net working capital. Quantum's net working capital more than doubled which was a sign of a rapidly growing company. Quantum's current ratio was undesirably low in 1991 because it was below 2. The current ratio strengthened considerably in 1992 to 2.63. Finally, its interest coverage was very healthy. Quantum's interest

Table 12.1. Financial Summary for Quantum[34] and Micropolis[35]

(dollar amounts in thousands)	Net sales		Net profits		Sustainable rate of growth	
	1992	1991	1992	1991	1992	1991
Quantum	$1,697,240	$1,127,732	$93,811	$46,845	23.6%	15.2%
Micropolis	$396,579	$350,875	$19,557	$4,343	14.3%	3.8%

	Net working capital		Current ratio		Interest coverage	
	1992	1991	1992	1991	1992	1991
Quantum	$513,786	$212,160	2.63	1.88	11.2	10.6
Micropolis	$163,394	$141,850	4.84	3.89	5.09	1.73

coverage was well over the three to four times coverage that analysts use as a benchmark.

Now let's take a look at Micropolis. First of all, it was a smaller company than Quantum. Micropolis' 1992 sales were about a quarter of Quantum's. Micropolis was not growing as quickly as Quantum. Micropolis' sales increased only 13% in 1992 compared to 1991, and its sustainable rate of growth was 14.3% in 1992. Also its working capital, although growing, grew only 15% in 1992.

But notice the jump in net profits. Even though sales increased by only 13%, net profits more than quadrupled. This was an indication that Micropolis was operating near its break even point in 1991. Recall that when operating near the break even point, any small change in volume or sales will result in disproportionately large percentage swings in profits. The large jump in sustainable rate of growth from 3.8% to 14.3% was another indication that the company was close to the break even point in 1991. Therefore although net profits more than quadrupled in 1992, it must be understood that most of this improvement was due to operating leverage and being near the break even point in 1991.

In terms of financial health, we have already seen that net working capital increased 15% in 1992. Micropolis' current ratio looked very

---

34. Quantum Corporation 1993 Annual Report and Form 10-K.

35. Micropolis Corporation 1992 Annual Report.

good at 4.84 in 1992 up from 3.89 in 1991. This was much higher than Quantum's current ratio. However when examining Micropolis' interest coverage, it was dangerously low at 1.73 in 1991. This meant that for every $1 of interest expense, there was only $1.73 of available income (from operations and other income) to meet the interest expense. This was another sign Micropolis' operations were close to break even in 1991. It is desirable to have an interest coverage greater than three or four. Once Micropolis' operations were more profitable in 1992, its interest coverage rose to a much safer 5.09.

In summary, both Quantum and Micropolis were in the very competitive disk drive business. Quantum was much larger and had a higher sustainable rate of growth. Larger demand for disk drives in 1992 enabled both companies to improve their financial health in areas where they were weak. Quantum was able to improve its current ratio, and Micropolis was able to improve its interest coverage. More concern may be placed with Micropolis in light of the intense price pressure in the disk drive market. Its numbers showed what wild profit fluctuations could result when operating near the break even point. The price pressure made operating on the positive side of break even all the more difficult. Engineers at both companies could affect finances most significantly by developing new products with performance, reliability, and service advantages, and by accelerating the product development cycle. This would increase gross profit margins, eventually leading to higher sustainable rates of growth.

2. FINANCIAL LINKAGE ANALYSIS

Quantum's financial linkage analysis has already been discussed. The analysis is shown in Figure 11.2. Recall that Quantum's sustainable rate of growth improved in 1992 because its operating expenses relative to sales decreased, and it was able to take advantage of increased financial leverage. Possible areas for improvement were to increase gross profit margin and to decrease inventory levels.

Now let's take a look at Micropolis. Its financial linkage analysis is shown in Figure 12.1. This shows that gross profit margin increased from 18.6% to 22.7% due to a decrease in cost of sales. This change was the most significant change on the whole sheet, and it can be seen how this led to the increase in sustainable rate of growth from 3.8% to 14.3%. The higher gross profit margin led to a higher operating profit

Figure 12.1. Financial Linkage Analysis for Micropolis Corporation[36,37]

All dollar figures in thousands of dollars

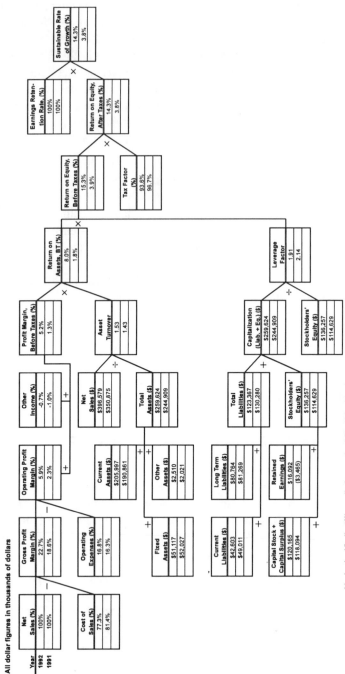

36. Lucien Virgile, "Micropolis Corporation," The Value Line Investment Survey, Apr. 30, 1993, 1101.
37. Micropolis Corporation 1992 Annual Report.

margin and a higher profit margin before taxes. Note how close Micropolis was to the break even point in 1991. Its profit margin before taxes was only 1.3%. Note how a 4 percentage point decrease in cost of sales led to a 4 percentage point increase in profit margin before taxes—going from 1.3% in 1991 to 5.2% in 1992.

After multiplying profit margin by a slightly improved asset turnover, the return on assets increased from 1.8% to 8.0%. The leverage factor actually decreased in 1992. Micropolis was trying to decrease its interest expense by decreasing the amount of funds borrowed in order to improve its interest coverage. Multiplying the leverage factor by the return on assets led to a return on equity before taxes of 15.3% in 1992 and 3.9% in 1991. The earnings retention rate for Micropolis was 100% since it paid no dividends to its stockholders. Thus its sustainable rate of growth was the same as the return on equity after taxes.

Financial linkage analysis shows Micropolis' improvement in sustainable rate of growth was due mainly to improved cost of sales. But how could the company's finances be improved further? Micropolis would have been hard pressed to increase its financial leverage because of the risks it faced with the price wars for its disk drive products and the prospect of operating near the break even point. Carrying increased debt would not have been prudent. Thus improvement in the company's finances would probably need to come from the operations side. In comparing Micropolis' numbers with Quantum's, it can be seen that operating expenses relative to net sales were 16.8% for Micropolis in 1992 compared to 10.2% for Quantum. Micropolis had an opportunity to improve its operating expenses. Another opportunity for improvement would be to continue utilizing its assets more efficiently. Micropolis' asset turnover improved from 1.43 to 1.53, but this was still less than Quantum's asset turnover of 1.83.

Chapter 13

CASE STUDY—MICROSOFT AND LOTUS DEVELOPMENT

M icrosoft's strategic plan as of 1986 was presented as a case study in Chapter 6. Chapter 13 continues this case study by analyzing the financial results. Recall that in 1986, Microsoft's goal was to overtake Lotus as the leading independent software supplier. It wanted to enhance its dominance in personal computer operating systems by introducing a new version of MS-DOS and by making Windows an industry standard. Microsoft could then take advantage of its intimate knowledge of the new operating systems and develop the best applications for them.

In the succeeding years, Microsoft's strategy worked to near perfection. In 1988, Microsoft was able to overtake Lotus in terms of net sales, and its financial performance continued to improve. In the five years of 1988 through 1992, Microsoft's net sales increased at an average annual rate of 48.5%, and its net profits increased at an average annual rate of 53.5%.[38] Meanwhile, Lotus' average annual growth rates for sales and profits increased at 27.5% and 13.0%, respectively, for the same time period.[39]

As of early 1993, the computer software industry was in the midst of a price war which had spurred sales, but also hurt the software companies' profits. As mentioned in the previous case study, personal

---

38. George A. Niemond, "Microsoft Corporation," *The Value Line Investment Survey,* Mar. 12, 1993, 2126.

39. George A. Niemond, "Lotus Development Corporation," *The Value Line Investment Survey,* Mar. 12, 1993, 2125.

computer manufacturers engaged in an all out price war in the early 1990's. Falling prices meant tough times for hardware manufacturers, but the less expensive equipment spurred buying. Strong unit sales of personal computers were a boon to software companies.[40] Undoubtedly Microsoft profited the most from this situation because it still dominated the personal computer, operating system market. Virtually all IBM compatible PCs still required Microsoft's DOS operating system, and Windows was entrenched as an industry standard. However other software makers were not willing to concede market share for applications software without a fight. Lotus had recently introduced an innovative spreadsheet called Improv at a special price of $99. This price was much lower than its price for Lotus 1-2-3. Microsoft had also been playing this game. In order to gain market share in the database market, it introduced its Access product at $99.

1. FINANCIAL SUMMARY AND FINANCIAL HEALTH ANALYSIS

Table 13.1 shows the pertinent data for this analysis. By all accounts, Microsoft appeared to be keeping up with its stunning growth of past years while maintaining very strong financial health. In comparing 1992 with 1991, Microsoft's sales increased by 50%, its net profits increased by 53%, and its net working capital increased by 80%. The sustainable rate of growth decreased slightly from 34.3% to 32.3%, but this result was still very strong. In terms of financial health, the current ratio increased to a very solid 3.96. And since Microsoft had no long-term debt, its interest payments consisted of payments to cover only current liabilities. Thus its interest payments were very low, and its interest coverage was a phenomenal 99.2.

In the case of Lotus, 1992 was a difficult year as shown in Table 13.1. Net sales increased by only 8.6%, and net profits decreased by 31%. To get a better idea of how the company's operations area generated profits, items that are normally non-recurring such as restructuring charges and an investment gain were not included in the net profit calculation. Sustainable rate of growth also decreased from

---

40.   George A. Niemond, "Computer Software and Services," *The Value Line Investment Survey,* Mar. 12, 1993, 2111.

Table 13.1. Financial Summary for Microsoft[41] and Lotus Development Corporation[42]

(dollar amounts in thousands)

	Net sales		Net profits		Sustainable rate of growth	
	1992	1991	1992	1991	1992	1991
Microsoft	$2,758,725	$1,843,432	$708,060	$462,743	32.3%	34.3%
Lotus	$900,149	$828,895	$45,697(a)	$66,116(b)	14.0%(a)	18.5%(b)

	Net working capital		Current ratio		Interest coverage	
	1992	1991	1992	1991	1992	1991
Microsoft	$1,322,759	$735,150	3.96	3.51	99.2(c)	41.7(c)
Lotus	$296,166	$207,670	2.36	1.89	7.34	6.65

Notes:

(a) Does not include restructuring charge, $15,000 and gain on sale of Sybase, $49,706

(b) Does not include restructuring charge, $23,000

(c) Interest coverage calculated assuming nonoperating expenses to be interest expense

18.5% in 1991 to 14.0% in 1992. In terms of financial health, the main concern for Lotus was the current ratio. Net working capital increased by a healthy 43%, and interest coverage remained fairly strong and actually improved in 1992. However the current ratio in 1991 was a concern being below 2. Financial linkage analysis will show that Lotus was able to improve this ratio by increasing current assets and decreasing current liabilities.

In short, by examining the financial summaries of the two companies, Microsoft appeared to be as healthy as a company could get, and its sustainable rate of growth was over 30%. The only concern was this rate of growth could not be sustained forever, and it was inevitable for the company to begin slowing down at some point. Lotus was concerned that its net profits were decreasing as well as its sustainable rate of growth. Its current ratio improved to 2.36 in 1992, but this was an area to be watched closely. The next section will offer clues as to how Lotus' financial position could be improved.

---

41.   Microsoft Corporation 1992 Annual Report.

42.   Lotus Development Corporation 1992 Annual Report.

2. FINANCIAL LINKAGE ANALYSIS

The financial linkage analysis for Microsoft is shown in Figure 13.1, and the analysis for Lotus is shown in Figure 13.2. At first glance, note how the operations of these software companies differed compared to the disk drive companies shown in the previous case study. For the software companies, gross margins were around 80% while for the disk drive companies, gross margins were about 20%. Operating expenses relative to sales ranged from 45% to 68% for the software companies and only 10% to 17% for the disk drive companies. These differences point out why it is difficult to make meaningful comparisons when performing financial analyses of companies in different industries. The most informative comparisons, and the best way to discover how key financial measures can be improved, result from examining competing companies in the same industry as well as comparing year to year results.

In analyzing Microsoft, the sustainable rate of growth decreased slightly from 34.3% to 32.3%, and financial linkage analysis shows why. Gross profit margin improved to an impressive 83.1% because cost of sales improved. However, operating expenses increased, so the result was only a slight improvement in operating profit margin and profit margin before taxes. The key factor in the decline of the sustainable rate of growth was asset turnover. This declined from 1.12 to 1.05. Thus even though sales grew rapidly, assets grew even faster. An examination of the balance sheet shows the asset that grew the most in 1992 was cash and short-term investments. This grew from $686 million in 1991 to $1.345 billion in 1992. Microsoft was growing so fast and it was so profitable that its cash was accumulating quickly. As the company grew, it was more difficult to invest the cash into areas that were as efficient at generating sales as previous assets. Thus asset utilization efficiency declined which resulted in a lower asset turnover. Microsoft was in an enviable position with its cash growing so rapidly, but this analysis points out why their rapid growth would be difficult to continue forever.

Because of lower asset turnover, the return on assets declined slightly from 40.8% to 39.4%. The bottom of Figure 13.1 shows that both total liabilities and stockholders' equity grew at about the same rate. Thus the leverage factor stayed almost the same. With no significant changes in the rest of the financial linkage, the lower return

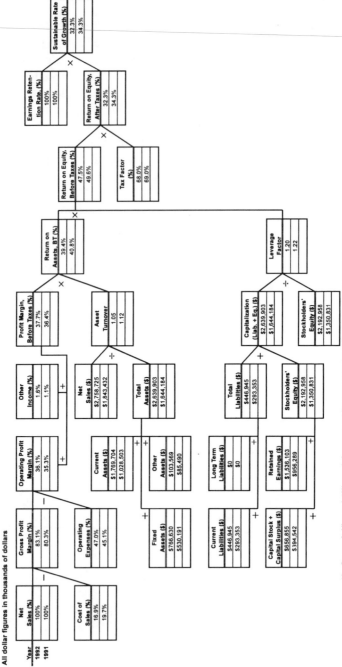

Figure 13.1. Financial Linkage Analysis for Microsoft Corporation[43,44]

All dollar figures in thousands of dollars

43. George A. Niemond, "Microsoft Corporation," The Value Line Investment Survey, Mar. 12, 1993, 2126.
44. Microsoft Corporation 1992 Annual Report.

Figure 13.2.   Financial Linkage Analysis for Lotus Development Corporation[45,46]

All dollar figures in thousands of dollars

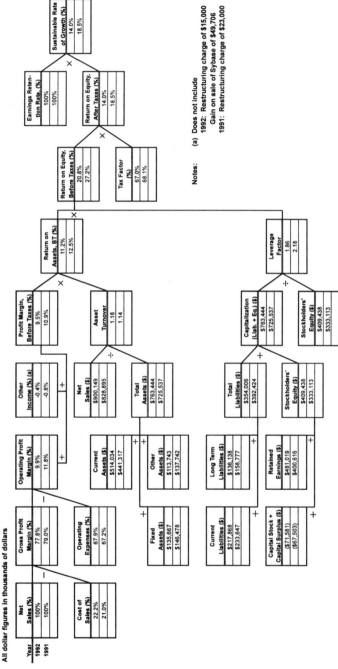

Notes: (a)   Does not include
1992: Restructuring charge of $15,000
Gain on sale of Sybase of $49,706
1991: Restructuring charge of $23,000

45.   George A. Niemond, "Lotus Development Corporation," The Value Line Investment Survey, Mar. 12, 1993, 2125.
46.   Lotus Development Corporation 1992 Annual Report.

on assets led to lower return on equity which led to a lower sustainable rate of growth.

In the case of Lotus, its operations area was much less profitable than Microsoft's. The operating profit margin in 1992 was 9.9% for Lotus and 36.1% for Microsoft. This was because Lotus' cost of sales and operating expenses relative to sales were much higher than Microsoft's. Also comparing 1992 to 1991, both cost of sales and operating expenses increased for Lotus, and this resulted in a decline in operating profit margin from 11.8% to 9.9%. Asset turnover increased slightly, but this was not enough to overcome the decline in profit margin. Therefore the return on total assets declined slightly from 12.5% to 11.2%.

Another area which contributed to a lower sustainable rate of growth for Lotus was the leverage factor. Recall that the company's current ratio in 1991 was a concern at 1.89. In order to improve the current ratio, current and long-term liabilities were driven down in 1992. Lowering liabilities improved its financial health but also led to a lower leverage factor. The lower leverage factor coupled with a lower return on total assets led to a lower return on equity before taxes of 20.8% in 1992 versus 27.2% in 1991. This ultimately resulted in the sustainable rate of growth declining from 18.5% to 14.0%.

In searching for areas where Lotus might improve, it was probably wise to avoid increasing the leverage factor again because this might have jeopardized the company's financial health. The area of focus would appear to be in the operations area. Efforts were needed to improve gross profit margins and operating expenses so operating profit margins would begin to approach those of Microsoft.

In summary, Microsoft had a phenomenal rate of growth. Its sustainable rate of growth was beginning to decline because its asset turnover was declining. The company was generating cash so quickly, it was difficult to invest this new cash in ventures that were as effective as previous ventures in generating sales and profits. This was an enviable position, but it pointed out the difficulties sustaining growth rates above 30%. Lotus on the other hand could have improved its sustainable rate of growth by improving its operations efficiency. It may not have been in a position to increase its leverage factor because it did not want to jeopardize its financial health.

Chapter 14
MAKING A DIFFERENCE AS AN ENGINEER

Some of the problems with studying the financial side of the business are (1) sometimes it is difficult to obtain the appropriate financial information, and (2) it is easy to get lost as to what the numbers really mean. The purpose of Part III is to provide an overview of finance. The key goals of this chapter are to show how you can get financial information and how you can affect these numbers. You must understand how your activities relate to the financial numbers. From previous chapters, it was shown how the sustainable rate of growth can be improved by lowering cost of sales, decreasing operating expenses, decreasing inventories, increasing financial leverage, etc. But when it comes down to wanting to make a positive impact, it can be very confusing as to what exactly to do or where to begin. This chapter will offer clues as to how to make the greatest financial impact.

One of these clues comes from the strategic plan. Recall in Part II on strategic planning that the main purpose of a strategic plan is to define the most important tasks for an organization to achieve long-term success. Another one of these clues comes from the financial goals, most notably the desired sustainable rate of growth. These two clues must be consistent and in agreement with one another. If they are not in agreement, THERE IS A PROBLEM which must be corrected by the executive management team.

For instance, there is a problem if a company desires to grow rapidly and have the leading market share of a particular business segment without the willingness or capacity to invest the money necessary to execute this plan. Another example is a company who wants to become the price leader for a particular product without a solid plan of how it intends to become the most efficient producer. In other words, there is a problem if the goals of the company, including financial goals, do not match the company's capabilities and business environment. With such inconsistencies, a firm can still be successful, but this success will come about because of luck instead of sound planning. With executives making sure the strategic plan is consistent with the financial goals, the chances of success are improved.

1. GETTING FINANCIAL INFORMATION

The most comprehensive way to gain an understanding of the financial situation for your organization is to look at it from both macroscopic and microscopic levels. Analyze at a macroscopic level by performing financial linkage analysis of your company and a competitor. Then analyze at a microscopic level by understanding the finances of your organization.

Performing financial linkage analysis requires getting balance sheets and income statements for your company and a competitor. These data can be obtained most easily if both companies are public. Public companies have their shares traded on various stock exchanges, and they are required to publish their finances on a quarterly basis.

Get the annual reports of both companies. One source of this financial information is over the Internet. Another method is to talk to your Finance department or Investor Relations department to get information about your company. For the competitor company, call up a stock brokerage company to either get the annual report or the phone number of the competitor. You can also get companies' phone numbers by consulting the reference section of your public library. When you call the company, ask for the Investor Relations department. Request a copy of the latest annual report. It is also beneficial to request a Form 10-K and a Form 10-Q. A Form 10-K is an annual report filed by public companies to the Securities and Exchange Commission which provides a comprehensive overview of the company's business including financial information as well as a description of the company's target market, business environment, risks, opportunities, and a description of key personnel. It must be filed within 90 days after the end of the company's fiscal year. A Form 10-Q is a report filed quarterly to the Securities and Exchange Commission, and it includes unaudited financial statements and provides a continuing view of the company's finances during the year. It must be filed within 45 days of the close of the first three fiscal quarters. You may also request the Investor Relations department to get on their mailing list if you want future quarterly and annual reports.

If the companies you wish to analyze are not public, getting this financial information is more difficult. Chances are you may not be able to get it. Ask people in your company's Finance department and at libraries for their ideas of getting financial information of the particular companies of interest. Search the Internet for press releases and any other information that may be available about these companies.

Whether or not you are successful at getting macroscopic information to perform your financial linkage analysis, the next step is to gain a microscopic level of understanding. Discuss with your manager and with people from the Finance department the finances for your organization or division. Determine the financial goals and the key measures used by them to measure financial performance. Then request data of these key financial measures for the last few years, so you can assess trends as well as the current financial status of your organization or division. Ask them for explanations of how your work activities can impact the key financial measures for your organization. Compare your understanding of your organization's strategic plans with its financial goals. Ask your manager to help clarify points that may seem inconsistent to you.

2. IMPROVING PRODUCTIVITY

When described in its most basic form, productivity is the relationship between output and input:

$$\text{Productivity} = \frac{\text{Output}}{\text{Input}} \tag{14.1}$$

Output can be defined as new products developed, sales, units produced, profits generated, etc. Input can be defined as the time and cost of buying and maintaining resources necessary to produce the output such as new product development costs, plant and equipment, personnel, etc. Improving productivity means increasing the ratio of output to input.

Most engineers know that a key contribution they can make is to improve productivity. The amount of success productivity improvements bring is largely within your control. For maximum benefit, productivity should be improved for tasks and areas deemed as crucial by the strategic plan and should help achieve the company's financial goals.

3. HIGH GROWTH VERSUS LOW GROWTH

In the ideal case, maximum productivity improvement is achieved by maximizing output and minimizing input. You should always strive to improve both output and input simultaneously. However in practice, it is sometimes difficult to affect changes in both simultaneously, and sometimes you must focus on one of these areas over the other.

In order to provide some general ideas where to focus your attention, let's look at a guideline of how resources are allocated depending on whether the company desires high growth or low growth. Table 14.1 summarizes tendencies for resource allocation and key financial measures for high growth and low growth organizations. Keep these in mind as you have discussions with your manager and Finance personnel about your work group's finances. This table is presented to serve as an overall guideline rather than absolute rules. High growth and low growth are relative terms and depend on the industry. There is also some overlap between high and low growth, and sometimes it is difficult to categorize your firm or organization as high or low growth. Thus the situation of your own firm must be analyzed independently, but Table 14.1 can provide clues as to where to focus your efforts.

The main point of Table 14.1 is that when inputs and outputs cannot be optimized simultaneously, successful high growth firms tend to invest in the business in order to maximize output, and successful low growth firms tend to increase efficiency in order to minimize input. For high growth firms, it is essential to recognize that investments need to be made in order to grow. As examples, these investments can be made into resources allocated to improve the following areas: new product development, manufacturing capacity, sales force expansion, and financial leverage. New products presumably stimulate growth and have higher gross margins than current products. Increasing manufacturing capacity allows the company to produce more products to sell. Expanding the sales force obviously is an attempt to get more

Table 14.1. Resource Allocation and Key Financial Measures for High and Low Growth Strategies

Growth Goal	Resource Allocation	Key Financial Measures
High growth (Invest to maximize output)	Develop new products Increase manufacturing capacity Expand sales force Increase financial leverage	Sustainable rate of growth Sales growth Return on investment Leverage factor Current ratio Interest coverage
Low growth (Increase efficiency to minimize input)	Improve margins Maintain/shrink product lines Maintain/shrink sales force	Sustainable rate of growth Gross profit margin Operating profit margin Asset turnover Return on assets

orders. Increasing financial leverage is an attempt to increase the sustainable rate of growth.

Of the methods to allocate resources for high growth, engineers have the most direct involvement with new product development and increasing manufacturing capacity. Any method engineers can utilize to speed up new product introductions is a great help to a growing company. Similarly, expanding manufacturing capacity by investing in new equipment or methods can be important for a growing company.

An example from personal experience is when the author worked on the development and manufacturing of a type of circuit from silicon wafers called a charge-coupled device (CCD). There was an eager customer for every good device that could be made. The problem was manufacturing enough CCDs to satisfy demand and to grow sales. Three barriers stood in the way. The manufacturing process was complicated, the yield of producing good devices was poor, and there was poor utilization of equipment because there was only one work shift.

The problems of a complicated manufacturing process and poor yield went hand in hand. Process steps originally introduced in an effort to improve manufacturing robustness were actually creating more defects which caused low yield. Such extraneous steps were identified and eliminated. This streamlined process flow led to a simpler process and resulted in higher yields.

The last problem of poor equipment utilization was alleviated by spending the money to start up a second work shift. The first work shift worked the day shift, and the second shift worked the swing shift. This doubled the equipment utilization.

Simplifying the production process and adding a second shift resulted in a tenfold increase in the production output of good CCDs. Allocating engineering resources to fix the manufacturing problems and spending the money to hire more workers led to a significant sales growth for the organization.

When examining the finances of a high growth company, expenses tend to be higher than for a low growth company. Investment in new fixed assets to expand production results in higher depreciation expenses which gets included in manufacturing overhead. This increases cost of sales and decreases gross profit margins. Also higher development costs and higher sales and marketing costs result in higher operating expenses. This leads to lower operating profit margins for a high growth company.

High growth companies are more likely to have newer, largely undepreciated fixed assets. Thus, net fixed assets are greater which leads to lower asset turnover. Lower asset turnover coupled with the lower profit margins leads to lower return on assets for a high growth company.

Recall from financial linkage analysis that return on equity is the product of return on assets and the leverage factor. Since a high growth company tends to have lower return on assets, financial leverage must be increased in order to boost return on equity. In practice this means financing the purchases of fixed assets with long-term debt. Thus high growth companies tend to rely more on long-term debt than low growth companies. Finally, in order to maximize the sustainable rate of growth, a high growth company tends not to pay dividends to its stockholders.

This means the earnings retention rate is 100% and all return on equity after taxes contributes directly to the sustainable rate of growth.

The key financial measures for a high growth company are listed in Table 14.1. The sustainable rate of growth is the primary measure. High growth companies want to maximize this as long as they can maintain their financial health. In order to increase the sustainable rate of growth, the leverage factor needs to be increased. However the current ratio and interest coverage need to be watched closely to make sure the company does not overextend itself. Sales growth in volume and dollars is important, as well as the return on investment of any investment made in the business.

In contrast, low growth companies focus on increasing efficiency in order to minimize their inputs. Methods of allocating resources include: improving margins, maintaining or shrinking product lines, and maintaining or shrinking the sales force. Improving margins by lowering cost of sales and operating expenses leads ultimately to a higher sustainable rate of growth. Maintaining or shrinking product lines down to the most profitable lines is one way of decreasing cost of sales and operating expenses. Maintaining or shrinking the sales force is a way of limiting operating expenses.

Of the resource allocation methods for low growth companies, engineers have the most direct control over improving margins by making products more efficiently. This should be the focus of engineers in a low growth company.

One of the ways the author contributed during his career involved improving gross profit margins for integrated circuit products. He worked with the Finance department to develop a cost model which incorporated each step of the manufacturing process. Once this cost model was constructed, it was easy to analyze and locate the major cost components. This analysis led to efforts to improve costs by improving product yield, packaging costs, and reducing test time costs. The cost model enabled him to quantify the benefit of each cost reduction project. After the cost reduction projects were completed, it was simple to use the product cost information along with information of the average selling price and quantity sold to calculate the impact the cost reduction efforts had on gross profits and gross profit margin. Through these efforts, the gross profit margin for the highest sales volume product in the product line increased from 30% to 67%. This type of result gets attention from management.

In a low growth company, investment in fixed assets is lower than high growth companies which leads to lower cost of sales. Therefore low growth companies tend to have higher gross profit margins. Also low growth companies have lower development costs and lower sales and marketing costs which leads to higher operating profit margins.

Low growth companies tend to have fixed assets that are largely depreciated which means they have lower net fixed assets. Thus asset turnover for a low growth company is higher. Higher asset turnover coupled with higher operating profit margins results in higher return on assets. Thus the operations area of a low growth company is more efficient than a high growth company.

Another area where low growth companies differ from high growth companies is financial leverage. Low growth companies do not make as much investment into fixed assets. They are generally trying to decrease long-term debt in order to decrease interest expenses and decrease risk. Thus low growth companies tend to have lower financial leverage.

Finally, low growth companies tend to have a lower earnings retention rate. Instead of reinvesting all their profits back into the business, they tend to share a portion of the profits with stockholders by distributing dividends.

The key financial measures for a low growth company are listed in Table 14.1. As with a high growth company, the sustainable rate of growth is a key measure. Even though the return on assets is higher for a low growth company, the leverage factor and earnings retention rate are lower. Thus the sustainable rate of growth may be lower for low growth companies. As mentioned previously, low growth companies focus on efficiency, and this means focusing on the operations. Thus the other key measures for low growth companies are key measures for the operations area: gross profit margin, operating profit margin, asset turnover, and return on assets.

In summing up how engineers can make a positive impact on the financial results, it must be understood that methods for improving productivity can differ depending on whether the company strives for high growth or low growth. For high growth, focus on increasing output by developing new products, improving the development cycle, and investing wisely to increase manufacturing capacity. For low growth, focus on decreasing input by increasing manufacturing efficiency.

Finally, in order to be effective at improving the product development cycle and manufacturing capacity and efficiency, engineers must utilize the tools of communication skills, statistics, and project management. These topics make up various parts of this book. Using effective communication skills in order to be aware of and focus on the most important goals, achievements, and problems was the topic of Part I. Statistics, used as the basis for technical decision making and continual process improvement, is the topic of Part IV. Another tool engineers must use is project management. This will be the topic of Part V. Part V describes how to determine return on investment. Any project, whether it involves new product development or manufacturing efficiency improvement, involves both costs and benefits. Each project needs to be planned and analyzed before it gets approved. Determining the project's return on investment is critical.

Chapter 15

INTERNATIONAL FINANCE

This chapter is included for those engineers who work for international companies or do business with companies in other countries. There is no doubt that the economies of the world are becoming increasingly dependent on one another. This means business firms are becoming more international. Engineers are becoming increasingly impacted by the effects of international finance in general and currency exchange rates in particular. When considering the purchase of equipment or materials to increase productivity for manufacturing, to reduce manufacturing costs, or to develop improved products, it is likely you are considering products from foreign firms as possibilities. This automatically makes you vulnerable to the effects of fluctuating currency exchange rates.

Recall from the model of the dynamic business that there are two areas of the business: the operations area and investment area. Figure 15.1 shows for a strictly domestic firm how each of these areas generate and use cash. The operations side generates cash by selling its goods and services, and it uses cash to pay for the cost of sales and its operating expenses. The investment side generates cash from short-term marketable securities or the sale of long-term debt instruments such as bonds. An example of costs incurred on the investment side are interest costs on these bonds.

On the other hand, when considering an international firm, there is an additional variable tying together cash inflow and cash outflow. This variable is the fluctuation of currency exchange rates, and this is shown in Figure 15.2. For an international firm, there is still cash inflow and outflow from the operations and investment areas, but the valuation of these flows depends on the exchange rates between the domestic currency and particular foreign currencies. For instance, the value of

158

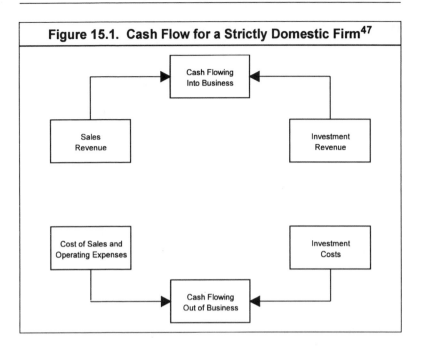

Figure 15.1. Cash Flow for a Strictly Domestic Firm[47]

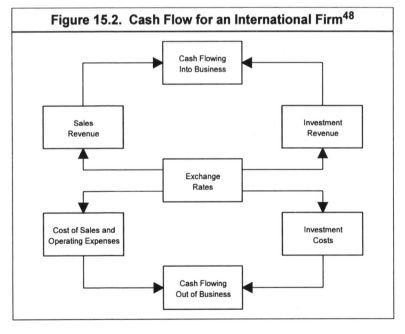

Figure 15.2. Cash Flow for an International Firm[48]

sales revenue depends on exchange rates for that portion of sales made to foreign countries (exports to foreign countries). The value of the cost of sales and operating expenses depends on exchange rates for that portion of goods and services from other countries (imports from foreign countries) used in order to produce or sell the product. Likewise, the valuation of investment revenues or costs made into or from foreign countries depend on currency exchange rates.

The purpose of this chapter is to provide an overview about what factors influence exchange rates and how exchange rate fluctuations can affect companies' finances.

1. FACTORS INFLUENCING EXCHANGE RATES

In the modern, global economy, many of the major countries have flexible exchange rates. Exchange rate is defined as the number of units of one currency required to purchase one unit of a different currency. These exchange rates are reported daily in major newspapers. See Table 15.1 as an example. The exchange rates can be expressed in U.S. dollars per unit of foreign currency or units of foreign currency per U.S. dollar.

Almost any newsworthy political or economic development can affect exchange rates in the short-term. However in the long-term, an important factor influencing exchange rates is the rate of inflation. Countries with high inflation rates will experience a reduction in the value of their currency relative to other countries. This will now be explained in more detail.

1.1. Inflation and the Purchasing Power Parity Condition

Opportunities for profits catch the attention of many people from around the world. One such opportunity that rarely goes unnoticed is the chance to buy something in one place and sell it in another place for a profit. Suppose for example the dollar price of gold in Tokyo was

---

47. Adapted from Maurice Levi, *International Finance: Financial Management and the International Economy,* New York: McGraw-Hill, 1983.

48. Ibid.

Table 15.1. Example of Exchange Rates for October 3, 1995[49]		
Country	U.S. $ per foreign currency ($/unit)	Foreign currency per U.S. $ (units/$)
Britain (Pound)	1.5830	0.6317
Canada (Dollar)	0.7524	1.3291
France (Franc)	0.2018	4.9565
Germany (Mark)	0.6957	1.4375
Italy (Lira)	0.0006186	1616.48
Japan (Yen)	0.009864	101.38

lower than the dollar price in New York at a particular time. If there is enough of a price difference, enough to cover transportation costs, people would buy gold in Tokyo and ship it to New York for sale. In general there are people who buy commodities in the cheaper market and sell them in the more expensive market. These people are known as commodity arbitrageurs.

Through the actions of commodity arbitrageurs, any profitable arbitrage opportunities are taken advantage of until the opportunity is eliminated. The arbitrageurs' actions raise prices in the low price markets, and decrease prices in the high price markets. With the exception of transportation costs and tariffs, prices in various locations become the same. The price of an individual commodity, in this case gold, will be equalized if the following relationship holds.

$$p_{US}^{gold} = S(\$/yen) \times p_{Japan}^{gold} \qquad (15.1)$$

In this equation, the p's are the gold prices in the United States in dollars and in Japan in terms of yen. $S(\$/yen)$ is the exchange rate of U.S. dollars per yen.

If Equation (15.1) is extended to hold for every item people buy, then it can be rewritten in terms of prices in general.

$$P_{US} = S(\$/yen) \times P_{Japan} \qquad (15.2)$$

where P_{US} and P_{Japan} represent prices in each respective country of a standard group of wholesale goods. This is the static form of the

---

49. *The Wall Street Journal,* Oct. 4, 1995.

purchasing power parity condition which is the law of one price. In practice, it will not hold perfectly because of differential transportation costs, differential tariffs on different products, and restrictions to free commodity arbitrage. However, this is a general expression to show how prices in different countries are related.

Now let's extend the purchasing power parity condition into its dynamic form. To do this, consider what this condition looks like a year into the future.

$$P_{US} \times (1 + \Delta P_{US}) = S(\$/\text{yen}) \times [1 + \Delta S(\$/\text{yen})] \times P_{Japan} \times (1 + \Delta P_{Japan}) \quad (15.3)$$

where $\Delta S(\$/\text{yen})$ is the percentage change in exchange rate over a year and ΔP_{US} and ΔP_{Japan} are the annual inflation rates in the United States and Japan. Now divide Equation (15.3) by Equation (15.2).

$$(1 + \Delta P_{US}) = [1 + \Delta S(\$/\text{yen})] \times (1 + \Delta P_{Japan}) \quad (15.4)$$

Rearranging Equation (15.4) to solve for $\Delta S(\$/\text{yen})$ gives

$$\Delta S(\$/\text{yen}) = \frac{\Delta P_{US} - \Delta P_{Japan}}{1 + \Delta P_{Japan}} \quad (15.5)$$

Assuming the foreign inflation rate is moderate, Equation (15.5) can be approximated by

$$\Delta S(\$/\text{yen}) \cong \Delta P_{US} - \Delta P_{Japan} \quad (15.6)$$

This is the dynamic form of the purchasing power parity condition. It states that change in exchange rates is approximately equal to the difference in inflation rates.

As an example, suppose the inflation rate in the United States is 5% ($\Delta P_{US} = 0.05$) and the inflation rate in Japan is 2% ($\Delta P_{Japan} = 0.02$). Then the dollar price of yen increases by about 3 percent.

$$\Delta S(\$/\text{yen}) \cong 0.05 - 0.02 = 0.03 \text{ or } 3\%$$

This means the yen is worth 3% more in terms of dollars, and in essence the dollar falls in value by 3% versus the yen. Thus the country with the higher inflation rate will experience a decrease in the value of its currency.

It must be pointed out that the purchasing power parity condition is at best a long-term approximation. There are many factors in the short-term that may influence exchange rates to deviate from the purchasing power parity condition. Such factors include political developments, tariffs and other barriers to free trade, interest rate differentials, and relative growth rates of national income. Influences from these factors mean years could pass before exchange rates move in accordance to the purchasing power parity condition.

In summary, the purchasing power parity condition is not followed strictly in all cases, but it gives you an idea where long-term exchange rates may be headed and an understanding of some factors affecting them.

2. THE EFFECT OF EXCHANGE RATE CHANGES ON INTERNATIONAL COMPETITIVENESS

This section will cover the first order effects exchange rate changes have on exporters and importers. Recall from Figure 15.2 that for an international company, exchange rates can affect the cash flowing into the business by changing valuation of its sales and investment revenue streams. Exchange rates can also affect the cash flowing out of the business by changing valuation of its cost of sales and operating expenses as well as investment costs. To take a closer look at these effects, consider the following three equations.

$$P_{US}(\$) = S(\$/yen) \times P_{Japan}(yen) \qquad (15.7)$$

$$P_{Japan}(yen) = S(yen/\$) \times P_{US}(\$) \qquad (15.8)$$

$$S(yen/\$) = \frac{1}{S(\$/yen)} \qquad (15.9)$$

Equation (15.7) states the price of an item in the U.S. in terms of dollars is the product of the Japanese price in yen multiplied by the exchange rate of dollars per yen. Equation (15.8) states the price of an item in Japan in terms of yen is simply the product of the U.S. price in dollars times the exchange rate of yen per dollar. The exchange rate of yen per dollar is just the inverse of the exchange rate of dollars per yen, as shown in Equation (15.9). The price of the item in Equations (15.7) and (15.8) can be for a revenue or cost component. This is an example of exchange rates between the U.S. and Japan, but other countries can be substituted in Equations (15.7) to (15.9).

2.1. Exporters

Let's assume in this simplified analysis of first order effects that a portion of the cash inflow, whether it is sales or investment revenue, comes from a foreign country and that all cash outflows are domestic. Thus we assume the exporter is not also an importer. So in order to examine the effect exchange rate changes have on an exporter, we only need to look at the impact on cash inflow.

Table 15.2 summarizes the effects of a devaluation of the U.S. dollar versus the Japanese yen. Consider a U.S. company that exports products to Japan. With a devaluation of the dollar, the exchange rate in terms of dollars per yen increases. Using Equation (15.7) and assuming the price of the product in Japan in terms of yen stays constant, the price of the exported product increases in dollar terms. This leads to higher profits in the short-term for the U.S. exporter. The company may even have the flexibility to decrease prices in Japan and still come out with higher dollar prices depending on the severity of the dollar devaluation. In the long run, the extra profits U.S. exporters enjoy will attract new competitors to the market which may tend to decrease profits somewhat from peak levels. This may limit the long-term extra profits the U.S. exporters receive.

On the other hand, considering a Japanese company exporting products to the U.S. shows the opposite effect. As shown in Equation (15.9), an increase in the exchange rate of dollars per yen equates to a decrease in exchange rate of yen per dollar. Using Equation (15.8), this means the price of the product in terms of yen decreases which leads to lower short-term profits for Japanese exporters.

Table 15.2. Effect of Exchange Rates on International Competitiveness

Example: The U.S. dollar devalues versus the Japanese yen, therefore
$S(\$/yen) \uparrow$
$S(yen/\$) \downarrow$

	Cash flow component	Type of company	Exchange rate	Revenue or cost component	Short-term effect
Exporter (Cash inflow)	Sales and Investment Revenue	U.S. exporter to Japan	$S(\$/yen)\uparrow$	$P_{US}(\$) \uparrow$	Profits $\uparrow$
		Japanese exporter to U.S.	$S(yen/\$)\downarrow$	$P_{Japan}(yen) \downarrow$	Profits $\downarrow$
Importer (Cash outflow)	Sales and Investment Costs	U.S. importer from Japan	$S(\$/yen)\uparrow$	$P_{US}(\$) \uparrow$	Profits $\downarrow$
		Japanese importer from U.S.	$S(yen/\$)\downarrow$	$P_{Japan}(yen) \downarrow$	Profits $\uparrow$

2.2. Importers

For this simplified analysis, assume that the importers have a portion of their cash outflow going to foreign countries and that all cash inflows are domestic. Thus we assume the importer is not also an exporter, so the examination of international competitiveness effects focuses only on the cash outflow effects.

Consider a U.S. company that imports products from Japan. This U.S. company uses the Japanese products to manufacture its own products which are sold in the U.S. The changing exchange rates has an effect on how much the imported products cost. A devaluation of the dollar in terms of yen means the cost of the Japanese products in terms of dollars will increase which leads to lower profits for the U.S. importer in the short run.

A Japanese importer will experience the opposite effect. A devaluation of the U.S. dollar means the yen is getting stronger and the price of the U.S. products in terms of yen in Japan decreases. With lower costs, the profits of the Japanese importer increases in the short-term. In the long-term, more competitors will be attracted to the market because of the extra profits. More competitors will serve to limit the amount of profits to be enjoyed by the Japanese importers.

The purpose of this section on international finance is to arm you with information that may affect your projects. If your home currency experiences a devaluation and your company is an exporter, short-term profits will tend to increase. But these extra profits should not create a false sense of security and cannot be relied upon in the long run. If your home currency experiences a devaluation and your company is an importer, costs will rise which place downward pressure on short-term profits. Thus if you are working on a cost reduction project, these extra costs need to be identified, separated out, and accounted for so the impact of your efforts are not misinterpreted by management.

Conversely, it is important to recognize the effects on international competitiveness when your home currency appreciates in value relative to other currencies. For exporters, short-term profits decrease, and for importers, short-term profits increase.

KEY POINTS: Finance

1) The organization can be viewed as a dynamic system, and finance is a way for the business to analyze its progress towards achieving its goals. Engineers should understand how financial planning fits with the strategic plan for the business.

2) The balance sheet and income statement are ways to measure financial performance. The balance sheet shows how strong the finances are by showing what the business owns and what it owes on a certain date. The income statement shows how the company did this year compared to last year by measuring profits and losses.

3) Financial linkage analysis can be used to get an overall picture how sales are tied to profits and ultimately to the sustainable rate of growth.

4) Engineers must understand how to make a difference. The key performance measures differ depending on whether the strategy is high growth or low growth. Maximize output for high growth, and minimize inputs for low growth. For engineers, maximizing output usually means developing new products, improving the product development cycle, and increasing manufacturing capacity. Minimizing inputs usually means improving manufacturing efficiency by lowering production costs.

5) As companies become more global, engineers should be more aware of world economic effects. Exchange rates and inflation rates are interdependent. Countries that have high inflation rates will eventually experience a reduction in the values of their currencies. Devaluations raise an exporter's profits and lower an importers profits in the short run. Conversely, countries that have low inflation rates compared with other countries will eventually experience an appreciation in the value of their currencies. Currency appreciation decreases profits for exporters and increases profits for importers.

SUGGESTED ACTIVITIES: Finance

1) Get financial information for your company and a competitor. Possible sources of this information are your company's Finance

or Investor Relations department, the Internet, stock brokerage offices, and the library. Request annual reports and Form 10-K information.

2) With the income statements and balance sheets obtained from Activity (1), perform financial linkage analysis for your company and a competitor by using the form shown in Figure 11.1. Look for differences in year-to-year results and differences with the competitor's finances. Identify components in the financial linkage analysis which will improve your company's sustainable rate of growth.

3) Learn of the key financial measures for your organization by asking your manager and personnel in the Finance department. Find out your organization's performance with respect to these key financial measures over the last few years.

4) Determine ways you can positively affect the key financial measures for your organization. Determine whether your organization emphasizes high growth and/or low growth strategies. Maximize outputs for high growth, and minimize inputs for low growth.

REFERENCES: Finance

General

Crainer, S. and Hodgson, P., *What Do High Performance Managers Really Do?*, Philadelphia: Trans-Atlantic Publications, Incorporated, 1993.

Downes, J. and Goodman, J. E., *Barron's Finance & Investment Handbook, 4th Edition*, Hauppauge: Barron's Educational Series, Incorporated, 1995.

Helfert, E., *Techniques of Financial Analysis: A Practical Guide to Managing & Measuring Business Practices, 8th Edition*, Burr Ridge: Irwin Professional Publishing, 1993.

Johnson, J. D. and Slottje, D. J., *Case Studies in Finance Using Excel*, New York: McGraw-Hill Companies, 1990.

Leahigh, D. J., *A Pocket Guide to Finance*, Orlando: Dryden Press, 1995.

Mayo, B., *Finance: An Introduction, 4th Edition*, Orlando: Dryden Press, 1992.

Milling, B. E., *The Basics of Finance: Financial Tools for Non-Financial Managers*, Naperville: Sourcebooks, Incorporated, 1991.

Peel, M., *Introduction to Management: A Guide to Better Business Performance*, Philadelphia: Trans-Atlantic Publications, Incorporated, 1993.

Pinches, G. E., *Essentials of Financial Management, 5th Edition*, New York: HarperCollins College, 1995.

Shimko, D. C., *Finance in Continuous Time: A Primer*, Cambridge: Blackwell Publishers, 1991.

Spiro, H. T., *Finance for the Nonfinancial Manager, 4th Edition*, New York: John Wiley & Sons, Incorporated, 1995.

Tracy, J. A., *The Fast Foward MBA in Finance*, New York: John Wiley & Sons, Incorporated, 1995.

Finding Financial Information

Howell, S., Editor, *Analyzing Your Competitor's Financial Strengths: How to Evaluate Key Threats & Opportunities*, Philadelphia: Trans-Atlantic Publications, Incorporated, 1993.

Lavin, M. R., *Business Information: How to Find It, How to Use It, 2nd Edition*, Phoenix: Oryx Press, 1992.

Financial Statements

Bandler, J., *How to Use Financial Statements: A Guide to Understanding the Numbers*, Burr Ridge: Irwin Professional Publishing, 1994.

Bernstein, L. A., *Analysis of Financial Statements, 4th Edition*, Burr Ridge: Irwin Professional Publishing, 1993.

Bukics, R. M., *Financial Statement Analysis: The Basics & Beyond*, Burr Ridge: Probus Publishing Company, Incorporated, 1991.

Costales, S. B., *Guide to Understanding Financial Statements, 2nd Edition*, New York: McGraw-Hill Companies, 1993.

Ferris, K. R., Haskins, M. E., and Selling, T. I., *International Financial Accounting & Reporting*, Burr Ridge: Richard D. Irwin, 1995.

Ferris, K. R., *How to Understand Financial Statements: A Nontechnical Guide for Financial Analysis*, Englewood Cliffs: Prentice Hall, 1992.

Kristy, J. E., *Analyzing Financial Statements: Quick & Clean, 5th Edition*, Buena Park: Books on Business, 1992.

Nyborg, R., *Analyzing Financial Statements Made Easy*, Provo: University Publishing House, Incorporated, 1994.

Pohlman, R., *Understanding the Bottom Line: Finance for Non-Financial Managers & Supervisors*, Overland Park: National Press Publications, 1991.

Tracy, J. A., *How to Read a Financial Report: Wringing Vital Signs Out of the Numbers, 4th Edition*, New York: John Wiley & Sons, Incorporated, 1993.

Weiss, D. H., *How to Read Financial Statements*, New York: A M A C O M, 1986.

Woelfel, C. J., *Financial Statement Analysis: The Investor's Self Study Guide to Interpreting & Analyzing*, Burr Ridge: Probus Publishing Company, Incorporated, 1993.

International Finance

El Kahal, S., *Introduction to International Business*, New York: McGraw-Hill Companies, 1994.

Gibson, H. D., *International Finance: Exchange Rates & Financial Flows in the International Financial System*, White Plains: Longman Publishing Group, 1996.

Gipson, C., *The McGraw-Hill Dictionary of International Trade & Finance*, New York: McGraw-Hill Companies, 1994.

Harris, J. M., Jr., *International Finance*, Hauppauge: Barron's Educational Series, Incorporated, 1992.

Holland, J., *International Financial Management, 2nd Edition*, Cambridge: Blackwell Publishers, 1993.

Levi, M., *International Finance: The Markets & Financial Management of Multinational Business, 2nd Edition*, New York: McGraw-Hill Companies, 1990.

McKinnon, R. I., *The Rules of the Game: International Money & Exchange Rates*, Cambridge: M I T Press, 1995.

McRae, T. W., *International Business Finance: An Introduction*, New York: John Wiley & Sons, Incorporated, 1995.

Melvin, M. H., *International Money & Finance, 2nd Edition*, New York: HarperCollins College, 1990.

Pool, J. C., *The ABCs of International Finance, 2nd Edition*, New York: Free Press, 1991.

Stein, Jerome L., *International Financial Markets: Integration, Efficiency & Expectations*, Cambridge: Blackwell Publishers, 1990.

Van der Ploeg, F., Editor, *The Handbook of International Macroeconomics*, Cambridge: Blackwell Publishers, 1993.

PART IV

STATISTICS

The Basis for Making Technical Decisions

Part III on finance showed a key financial measure for any business is the sustainable rate of growth, and engineers can improve this measure by improving the product development cycle and manufacturing efficiency. Statistics provides one of the tools for you to make such improvements.

Problems for engineers and engineering students associated with using statistics for Statistical Process Control and the Design of Experiments are as follows.

1) **Statistics is too theoretical.**
   Engineers and engineering students usually are only interested in statistics to the extent it will help their projects. However, courses in statistics at the university level are mostly theoretical. These courses also tend to be for a varied audience. Thus it is difficult for engineers to grasp practical applications of statistics for industrial and research settings. The purpose of Part IV is to focus on practical applications of statistics for engineers and engineering students.

2) **Management applies pressure for short-term results.**
   The most common course of action for management is to "get tough" whenever a technical problem arises. Management applies pressure on engineering personnel to do better and demands almost daily progress. This forces a one-variable-at-a-time experiment strategy and decisions made based on insufficient data. The use of statistics will help identify the problems that are statistically significant. Statistics will also lead to designing experiments that are more effective and efficient than one-variable-at-a-time experiments. Statistics can help pave the way to show management that a "get smart" philosophy should take the place of their "get tough" philosophy.

3) **Not enough time is spent on planning for data gathering and experimentation.**
   Engineers and engineering students are typically well trained at how to gather data and the mechanics of performing an experiment, but due to day-to-day pressures, not enough time is devoted to planning. Poor planning can lead to incorrect decisions being made about the data. Part IV will present strategies for planning effective control charts and experiments.

Statistics provides a set of tools which forms the basis for technical decision making. Its methods are a way of stating facts in terms of numbers. And even though the theory is based on advanced mathematics, most of the applications require only simple arithmetic. These tools will help identify and solve problems. For engineers, practical application of statistics can improve the results you get to develop new products and to improve manufacturing efficiency. For graduate students, these tools can improve the progress of your research projects. And for undergraduate students, these concepts are important to keep in mind while performing your laboratory experiments and to utilize throughout your career.

Part IV is intended for those of you who have a periodic need to use statistical techniques but don't have the time or inclination to become an expert on statistical theory. It begins with a brief overview of statistics and a comparison between the topics of Statistical Process Control (SPC) and Design of Experiments (DOE). The next chapter focuses on Statistical Process Control—the main tool to identify problems. The final chapter of Part IV describes Design of Experiments—the main tool to solve problems. Examples will be presented throughout Part IV to make the material more understandable and to demonstrate practical applications.

Chapter 16

OVERVIEW OF STATISTICS

Using statistics is a way for you to better perceive and understand the true state of nature. The true state of nature may never be understood completely, but statistics provides a set of tools to make our understanding more precise by use of hypotheses, theories, assumptions, and confidence levels.

1. THE TRUE STATE OF NATURE

There are certain mathematical relationships that exist between various elements in the true state of nature. These relationships can be straightforward. But most of these relationships are very complicated and may never be fully understood. Engineers can only try to perceive the true state of nature through experimentation. Experimental results will not exactly match the true state of nature due to experimental tolerances. For instance, suppose the actual life of an integrated circuit in the true state of nature is defined by the following equation.

Life of Integrated Circuit = f(size of transistors, number of transistors, operating temperature, operating voltage, operating current, thickness of layers in transistor, types of layers in transistor, cleanliness of manufacturing process)

Each of these relationships may be simple or complex. For example, suppose the actual f(operating temperature) portion of the equation is:

Life of Integrated Circuit $= 5.43e^{\frac{-2.62}{\text{Temperature}}}$

Suppose an engineer is given the task to increase the life of an integrated circuit. Obviously this task would be easy if the actual equation is known. However in practice, this equation is not known. Thus the engineer needs to simplify the investigation and devise a series of tests in an attempt to quantify the relationship between integrated circuit lifetime and other factors such as operating temperature and voltage. The actual equation will not be determined because results vary from experiment to experiment. For example, there will be slight variations in factors that are attempted to be held constant. Also there can exist functions unknown to the engineer that are part of the actual equation. The difference between experimental results and the true state of nature is illustrated in Figure 16.1.

Thus testing can only be used to estimate the true state of nature. Statistics are tools that help the engineer make these estimates more precise.

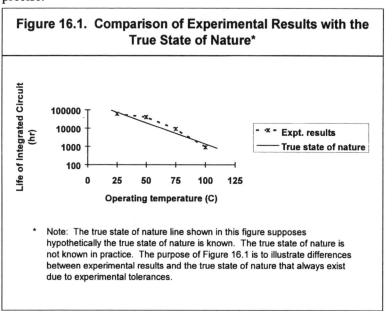

Figure 16.1. Comparison of Experimental Results with the True State of Nature*

*   Note: The true state of nature line shown in this figure supposes hypothetically the true state of nature is known. The true state of nature is not known in practice. The purpose of Figure 16.1 is to illustrate differences between experimental results and the true state of nature that always exist due to experimental tolerances.

2. DEFINITIONS

2.1. Populations

A population is a group of items upon which a decision is to be made. For example if it is desired to obtain yesterday's average yield of sucrose in a sugar refining process, the population would consist of all batches of sucrose produced yesterday. In research and development, populations are defined a little differently. For example consider trying to improve the tensile strength of a material by blending in a fiber additive. There are two populations. One population is all the material that would be produced if the fiber additive is not blended in, and another population is all the material that would be produced if the fiber additive is blended in.

In order to use statistics to provide an estimate of the entire population, samples of the population are taken. Statistics provides the method to infer population behavior based on a known sample. It is important that the sample comes from the population about which a decision is to be made and that it is representative of the population.

2.2. Distributions

The distribution is the shape of the data when plotted in a histogram. An example of a histogram is shown in Figure 16.2. A histogram plots the number of times an event occurs within specified intervals. For instance, assume the data in Figure 16.2 represent measurements taken from samples of a production process. From these samples, there are 8 occurrences of measurements between 20 and 30, 20 occurrences of measurements between 30 and 40, and so on.

There are many types of distributions such as uniform, binomial, Poisson, and lognormal. But engineers are most familiar with the

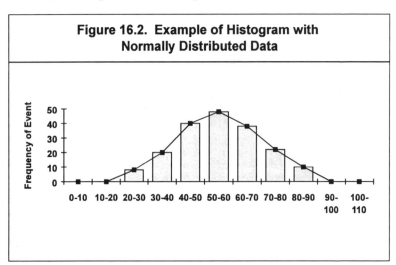

Figure 16.2. Example of Histogram with Normally Distributed Data

normal distribution. A normal distribution has the following characteristics:

- it has a single peak

- it is symmetrical on either side of the peak

- the distribution drops rapidly to zero away from the peak.

In Part IV, decisions will be made about sample means taken from the population rather than individual items taken from the population. There is a useful phenomenon pertaining to sample size and the central limit theorem. The central limit theorem states that as long as the sample size is large enough, the sample means will be distributed normally about the population mean regardless of the population distribution. Thus even though the distribution of the population may be unknown or it may not be normally distributed, it is still possible to use equations presented here to design sound experiments. As long as the sample size is sufficient, the sample means will be distributed normally about the population mean. This phenomenon is a critical element which simplifies the mathematics presented in Part IV. It also points out working with adequate sample sizes is crucial.

As an example of how statistics will be used to facilitate decision making, let's go back to the data presented in Figure 16.2. Suppose at a later date five more samples are taken from the production process, and all five samples measure between 80 and 90. It is highly unlikely that the process during the second sampling is the same as the process during the first sampling. Statistics will be used to help determine when changes have taken place.

2.3. Means and Variances

Statistics is used to interpret sets of data. Under most engineering situations, once a set of data is obtained under a fixed set of conditions, the data can be defined by three numbers:

- the number of samples (N)

- the mean of the samples ($\overline{X}$)

- the variance of the samples (S^2).

For N number of test results, the sample mean ($\overline{X}$) is defined as:

$$\overline{X} = \frac{\sum\limits_{i=1}^{N} X_i}{N} \tag{16.1}$$

The sample mean is a measure of the central tendency of the data.

The sample variance (S^2) is:

$$S^2 = \frac{\sum\limits_{i=1}^{N} \left(X_i - \overline{X}\right)^2}{N - 1} \tag{16.2}$$

The sample variance (S^2) is the square of the sample standard deviation (S) and both values are a measure of scatter of the data. The larger the scatter of data about the sample mean, the larger the variance and standard deviation. Note the denominator of Equation (16.2). The denominator $(N\text{-}1)$ is called the degrees of freedom and is denoted by the Greek letter phi (ϕ).

$$\phi = N - 1 \tag{16.3}$$

An important distinction must be made between statistics of a sample and of a population. When controlling a process or conducting an experiment, the engineer tries to find information about a certain population. Most often it is impossible, impractical, or uneconomical to test the entire population. So the engineer compromises by selecting and testing a number of items chosen to represent the entire population. This group of items is known as a random sample.

A sample mean is only a single estimate of the population mean. The population mean, referred to earlier as the true state of nature, is usually unknown to the engineer. The population mean is designated by the Greek letter mu (μ).

A sample variance is only a single estimate of the population variance. The population variance is denoted by the square of the Greek letter sigma (σ^2), and the population standard deviation is denoted by sigma (σ). It is common to hear about data that fall within certain multiples of sigma from the mean. As shown in Figure 16.3, 68.3% of the population falls within $\pm 1\sigma$, 95.4% falls within $\pm 2\sigma$, and 99.7% falls within $\pm 3\sigma$ of the mean for a normal distribution. In practice, it is

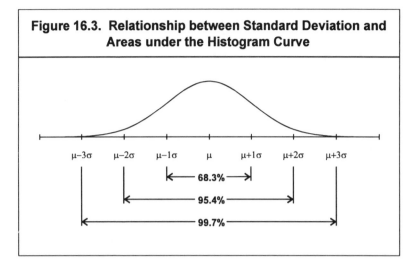

Figure 16.3.   Relationship between Standard Deviation and Areas under the Histogram Curve

common to represent the common variation as ±3σ since almost all the data fall within this range.

To summarize key points about statistics definitions, a random sample taken from a population can be characterized by three numbers:

- the number of items in the sample (N)

- the sample mean ($\overline{X}$)

- and the sample variance (S^2).

The sample mean ($\overline{X}$) is not equal to the population mean (μ). The sample mean is only a single estimate of the population mean. And the sample variance (S^2) is not equal to the population variance (σ^2). The sample variance is only a single estimate of the population variance.

3. SIGNALS VERSUS NOISE

Statistics can be viewed as a method for distinguishing signals from noise. Recall again the example of measurements taken from a production process that were shown in Figure 16.2. These data are shown again in Figure 16.4. There are variations around the mean called the variance. The variance can be the result of unexplained variations due to small differences in the product or in the method of testing. This variation is the noise of the process.

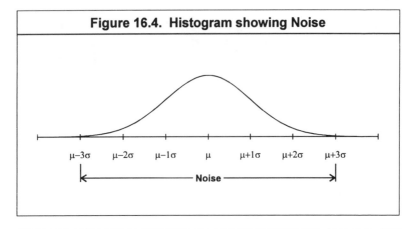

Figure 16.4. Histogram showing Noise

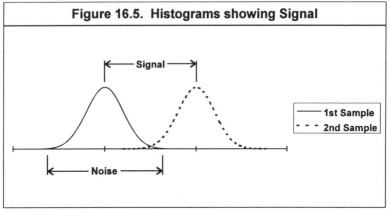

Figure 16.5. Histograms showing Signal

However suppose another sample is measured at a later date. This second sample has its own distribution that is shown in Figure 16.5. Because of the large differences in the means relative to the variances, it is highly unlikely the population from which the second sample was drawn is the same as the population from which the first sample was drawn. The change in means is a signal. The process has changed in some way.

Statistics will be used to determine whether variation in results can be explained by a nonrandom factor (assignable cause) or whether the variation is due to the inherent noise (common cause). Both Statistical Process Control and Design of Experiments utilize rules to judge signal-to-noise ratios to determine the likelihood of variations coming from significant changes or from background noise.

Chapter 17

SPC VERSUS DOE

Before delving into the details of SPC and DOE, this chapter relates how both of these methods identify signals from noise and their differences.

Take a look at Figure 17.1. Both of these methods attempt to distinguish signals from noise. In the case of SPC, random samples are taken from a production process over time. Variations about a mean are expected and are called common causes. The purpose of SPC is to identify significant changes. These changes are called assignable causes. When assignable causes exist, it signals you to examine the process more closely in order to find explanations for the change. SPC is passive from the standpoint that data are collected over time until a significant change is identified.

On the other hand, DOE is active because experiments are designed to cause significant changes. Once a problem has been identified, DOE is used to try and solve the problem. As seen in Figure 17.1, DOE results in data that are distributed around a mean similar to SPC. The distribution spread is due to slight variations in the product or measuring technique and is labeled experimental error. The signals are significant differences caused by varying certain parameters of the process.

Table 17.1 provides a summary comparing SPC and DOE. The objective of SPC is to identify problems, while the objective of DOE is to solve problems. Using SPC, engineers try to detect significant changes which signal assignable causes exist. On the other hand with DOE, engineers purposefully try to affect significant differences so that this knowledge can be incorporated into the process in order to improve it.

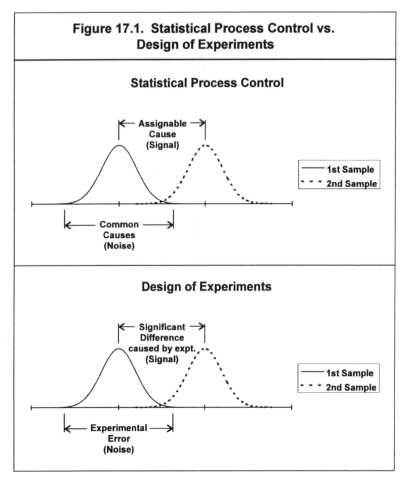

Figure 17.1. Statistical Process Control vs. Design of Experiments

In terms of sampling, both methods require careful planning. Planning and thinking ahead were shown to be crucial elements in Part I when developing a personal strategic plan and preparing for communications with others. Planning is essential for businesses when developing strategic plans (as described in Part II). Planning is also essential for the implementation of SPC and DOE. For SPC, samples are drawn from production over time. Whereas for DOE, samples are taken from experiments.

Analytical tools are different for the two methods. SPC uses control charts. Estimates of noise are made from the range or standard deviation chart. This estimate of noise is then used to calculate control limits for

Table 17.1. Comparison between Statistical Process Control and Design of Experiments[50]

	SPC	DOE
Objective	Detect significant differences	Detect significant differences
	Watch for assignable causes	Put assignable causes into process
	Problem identification	Problem solving
Sampling	Random samples carefully planned over time	Randomized by carefully planned experiment
Analytical tools	Control charts	Matrix designs
	Estimate of common cause (noise) from R or S chart	Estimate of common cause (noise) from SPC or experimental error
	Compare means against noise (control limits)	Compare means against noise (test criterion)
Uses	Controlling production	Initial development
		Setting up production
		Troubleshooting

the means chart. Sample means are then compared with the control limits. If the control limits are exceeded, this signals a significant change, and search for an assignable cause should begin.

DOE uses a tool called matrix designs. This helps the experimenter extract the maximum amount of information from the minimum number of samples. The estimate of noise comes either from prior knowledge such as from a control chart or comes from the results of the experiment itself. From this estimate of noise, a test criterion value is computed. Sample means from the experiment are compared with the test criterion value. A significant difference occurs when the test criterion value is exceeded.

Uses for DOE include characterization of a product or process during the initial development stages, setting up the product or process for production, and to troubleshoot problems. SPC is used mainly to monitor and control production. You should be familiar with both SPC and DOE so you can apply these tools as dictated by the nature of your job, research project, or experiment.

50.   Adapted from Don F. Wilson, *Design of Experiments Utilizing Taguchi Concepts,* Littleton, CO: QINAS, Inc., 1986.

Chapter 18

STATISTICAL PROCESS CONTROL

SPC utilizes control charts. An example of a control chart is shown in Figure 18.1. This is an example of an $\overline{X}$, R chart where $\overline{X}$ represents the sample means and R represents the sample ranges. Plotting the sample range is a way of determining the common cause variation, and it is used when the sample sizes are too small to calculate sample standard deviations.

Information from the R chart is used to calculate the upper and lower control limits for the $\overline{X}$ chart. In general, sample means that fall within the control limits represent common cause variation. Sample means outside the control limits signal a change in the process and an assignable cause exists. Using control charts is a simple way to display and analyze production data.

1. STEPS FOR CONSTRUCTING AND USING CONTROL CHARTS

The steps for initiating a control chart are outlined in Table 18.1. The most important aspect for successful implementation of SPC is careful planning. Planning for using control charts involves stating the objective and determining the subgroups. Some possible objectives are:

- To analyze a process to see if certain specifications can be met.

- To identify and eliminate assignable causes of variation.

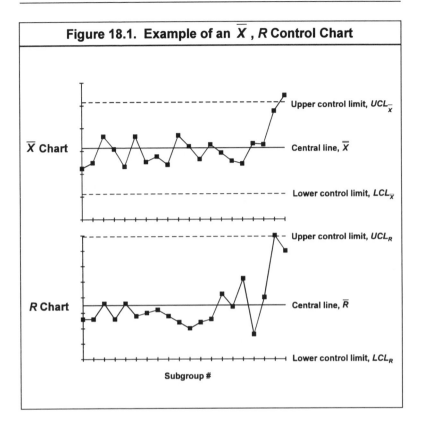

Figure 18.1. Example of an $\overline{X}$, R Control Chart

**Table 18.1. Outline for Constructing
and Using Control Charts**

1. State the objective

2. Determine the subgroupings

3. Collect the measurement data and calculate the sample
   statistics for each subgroup

4. Calculate the control limits

5. Plot the data

6. Interpret the results

7. Take action based on the results

8. Continue use of the charts

- To provide a basis for acceptance or rejection of manufactured product.

After the objective is defined, decide what will be the parameter to be measured and in what subgroups. The parameter must be measurable and expressed in terms of numbers. There are two primary types of parameters for control charts: variable parameters and attribute parameters. Variable parameters are actual measured quality characteristics such as weight, voltage, and length. X, R and X, S control charts are used for variable parameters. Attribute parameters are measures of go, no-go tests and are expressed in terms of fraction conforming or non-conforming. Examples of attribute parameters are the percentage of good product during testing or the fraction of reject product during inspection. When working with attribute parameters, p control charts are used.

The key idea when measuring parameters is to divide the measurements into rational subgroups. A subgroup is a sample you will measure to estimate statistics of the population. Each subgroup should be as homogeneous as practically possible while trying to maximize the opportunity for variation between subgroups. The sample size of each subgroup should be at least four in order for the central limit theorem to be in effect. This means the distribution of sample means will be nearly normal even though samples may be taken from a non-normal population.

As a rule of thumb, choose larger sample sizes to detect smaller variations in sample means. If working with variable data, use sample sizes of 10 or 20 when detection of small variations in means is desired. If the sample size is less than 15, the range chart should be used. If the sample size is 15 or greater, the S chart should be used. Use the p chart for attribute data no matter the sample size.

Once the subgroupings have been determined, begin collecting measurement data and compute the sample statistics for each subgroup. The type of statistics to be calculated depends on the type of control chart. In general, it is desirable to collect data from about 20 subgroups before control limits are calculated. After the data have been collected, calculate the control limits, and plot the data. The next step is to interpret the initial data. The final step is to continue using the control chart, recalculating control limits when deemed appropriate. The next sections describe the various types of control charts in more detail.

2. $\overline{X}$, R CONTROL CHARTS

$\overline{X}$, R control charts are used when the data are variable data and the subgroup sample size is less than 15.

2.1. Subgroup Calculations

After data have been taken from about 20 subgroups, there are two statistics to calculate from each subgroup: the sample mean ($\overline{X}$) and the sample range (R). The sample mean was shown in Equation (16.1).

$$\overline{X} = \frac{\sum\limits_{i=1}^{N} X_i}{N} \tag{16.1}$$

The sample range is shown in Equation (18.1).

$$R = X_{max} - X_{min} \tag{18.1}$$

where X_{max} is the maximum value of X within the sample and X_{min} is the minimum value of X within the sample.

2.2. Control Limit Calculations

For the case of $\overline{X}$, R charts, the population standard deviation (σ) is not known. The sample range data are used to estimate the population standard deviation. To make this estimate, an additional element of uncertainty is present in the decision making process. This additional uncertainty is accounted for by using the t distribution rather than the normal distribution. The t distribution has the same shape as the normal distribution, but the spread is wider. The width of the t distribution is a function of the degrees of freedom (ϕ). The degrees of freedom was shown in Equation (16.3) to be related to the sample size: $\phi = N-1$. Figure 18.2 shows a comparison of the t distribution and the normal distribution. Points to be noted from Figure 18.2 are:

- ϕ is a measure of the precision with which σ has been estimated.

- As ϕ increases, the width of the t distribution decreases.

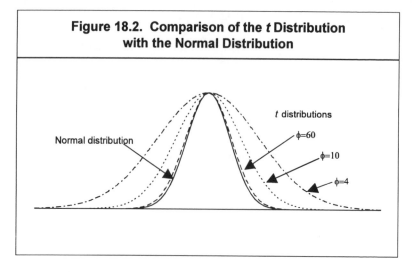

**Figure 18.2. Comparison of the *t* Distribution
with the Normal Distribution**

- When ϕ reaches about 60, the *t* distribution closely
  approximates the normal distribution, and σ is known for
  all practical purposes.

Thus when using R to estimate σ, the estimate will vary depending on
the sample size.

To calculate the control limits, take the sample means and sample
ranges that were calculated for each subgroup and compute the mean
of the sample means ($\overline{\overline{X}}$) and the mean of the sample ranges ($\overline{R}$).

$$\overline{\overline{X}} = \frac{\sum \overline{X}}{n} \tag{18.2}$$

$$\overline{R} = \frac{\sum R}{n} \tag{18.3}$$

where n is the number of subgroups.

Then use Table 18.2 to calculate the upper and lower control limits
for the $\overline{X}$ and R charts. The control limits represent the 3 sigma limits
for the distribution of $\overline{X}$ values.

For the $\overline{X}$ chart:

Table 18.2. Control Limit Calculations for X, R Control Charts

$\overline{X}$ chart	R chart
Central line = $\overline{\overline{X}}$	Central line = $\overline{R}$
$UCL_{\overline{X}} = \overline{\overline{X}} + A_2\overline{R}$	$UCL_R = D_4\overline{R}$
$LCL_{\overline{X}} = \overline{\overline{X}} - A_2\overline{R}$	$LCL_R = D_3\overline{R}$

Number of observations in subgroup	A_2	D_3	D_4
2	1.88	0	3.27
3	1.02	0	2.57
4	0.73	0	2.28
5	0.58	0	2.11
6	0.48	0	2.00
7	0.42	0.08	1.92
8	0.37	0.14	1.86
9	0.34	0.18	1.82
10	0.31	0.22	1.78
11	0.29	0.26	1.74
12	0.27	0.28	1.72
13	0.25	0.31	1.69
14	0.24	0.33	1.67
15	0.22	0.35	1.65

$$\text{Upper Control Limit}_{\overline{X}} = UCL_{\overline{X}} = \overline{\overline{X}} + 3\sigma_{\overline{X}} \qquad (18.4)$$

$$\text{Lower Control Limit}_{\overline{X}} = LCL_{\overline{X}} = \overline{\overline{X}} - 3\sigma_{\overline{X}} \qquad (18.5)$$

The factor $A_2\overline{R}$ from Table 18.2 is used to estimate $3\sigma_{\overline{X}}$. Table 18.2 shows A_2 decreases as the number of observations in the subgroups increases. Thus as the number of observations increases, the estimate of $3\sigma_{\overline{X}}$ becomes more precise.

For the R chart:

$$\text{Upper Control Limit}_R \ = \ UCL_R \ = \ \overline{R} + 3\sigma_R \qquad (18.6)$$

$$\text{Lower Control Limit}_R \ = \ LCL_R \ = \ \overline{R} - 3\sigma_R \qquad (18.7)$$

The factor $D_4 \overline{R}$ is used to estimate the upper limit for the 3 sigma variation of R values, and $D_3 \overline{R}$ is used to estimate the lower limit. Again as shown in Table 18.2, the precision of this estimate depends on the size of the subgroup.

Now that some of the mechanics of the $\overline{X}$, R control charts have been explained, let's look at an example.

2.3. $\overline{X}$, R Control Chart Example

At a salmon cannery, it is essential the amount of product in the cans meets the label specification of 227 grams (8 ounces). It is also desirable to ensure too much product does not get placed into the cans. Excess product in cans increases cost and also makes it more difficult to achieve the proper vacuum in the can. This upper specification limit is 257 grams. The target is 242 grams. To fill the cans, slices of salmon are injected into the cans by a filler machine. The filler machine has two heads that alternate filling cans in order to double the throughput. While one head is injecting salmon into a can, the other head gathers salmon for the next can. When the next can comes through, the second head injects its salmon while the first head gathers salmon for the following can.

The salmon cannery has three canning lines. Each line has a separate filler machine.

- **Objective**

  The primary objective is to determine whether or not the process is in control and to minimize variations by identifying assignable causes. Another objective is to determine if the amount of product in the cans meets the specification on the label.

- **Determine subgroupings**

  A separate control chart should be kept for each of the three production lines since each line has a different filler machine.

This will ensure that each subgroup is homogeneous. This example will concentrate on the data from one of the production lines.

To break up the subgroups further, one possibility is to track data separately from each of the two heads of each filler. However for this case, it was determined that this amount of detail would be too difficult to achieve in practice.

The production lines run continuously for 10 hours per day. Thus a rational subgroup would be to take samples on a time basis such as every hour. A sample size of five cans for each subgroup would be easily doable.

Conclusion: the subgroupings will consist of five cans to be taken from each production line every hour. Each of these cans will be weighed to determine the amount of product in them.

- **Collect measurement data and calculate $\overline{X}$ and R for each subgroup.**

Time	1	2	3	4	5	$\overline{X}$	R
7/8, 8:00 a.m.	238	243	235	230	235	236.2	13.0
9:00 a.m.	242	235	229	240	241	237.4	13.0
10:00 a.m.	250	239	252	241	234	243.2	18.0
11:00 a.m.	235	234	247	245	241	240.4	13.0
12:00 p.m.	231	246	243	228	235	236.6	18.0
1:00 p.m.	260	246	230	240	240	243.2	30.0
2:00 p.m.	237	230	245	233	243	237.6	15.0
3:00 p.m.	229	245	247	240	233	238.8	18.0
4:00 p.m.	234	238	239	230	244	237.0	14.0
5:00 p.m.	250	245	243	241	238	243.4	12.0
7/9, 8:00 a.m.	243	239	239	241	243	241.0	4.0
9:00 a.m.	243	243	231	236	238	238.2	12.0
10:00 a.m.	246	247	234	236	244	241.4	13.0
11:00 a.m.	246	235	230	236	251	239.6	21.0
12:00 p.m.	243	247	235	234	230	237.8	17.0
1:00 p.m.	238	230	231	231	256	237.2	26.0
2:00 p.m.	238	239	242	246	243	241.6	8.0
3:00 p.m.	247	235	235	255	235	241.4	20.0
4:00 p.m.	230	265	236	243	270	248.8	40.0
5:00 p.m.	262	235	269	261	234	252.2	35.0
					Mean	240.7	18.0

- **Calculate control limits**

$$\overline{\overline{X}} = 240.7$$

$$\overline{R} = 18.0$$

$$UCL_{\overline{X}} = \overline{\overline{X}} + A_2\overline{R}$$

   $A_2 = 0.58$ for subgroup sample size of 5 (from Table 18.2)

$$UCL_{\overline{X}} = 240.7 + (0.58)(18.0) = 251.1$$

$$LCL_{\overline{X}} = \overline{\overline{X}} - A_2\overline{R} = 240.7 - (0.58)(18.0) = 230.3$$

$$UCL_R = D_4\overline{R}$$

   $D_4 = 2.11$ for subgroup sample size of 5 (from Table 18.2)

$$UCL_R = (2.11)(18.0) = 38.0$$

$$LCL_R = D_3\overline{R}$$

   $D_3 = 0$ for subgroup sample size of 5 (from Table 18.2)

$$LCL_R = (0)(18.0) = 0$$

- **Plot the data**

$\overline{X}$ Chart: Salmon Can Weights

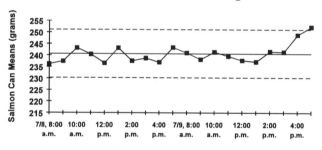

R Chart: Salmon Can Weights

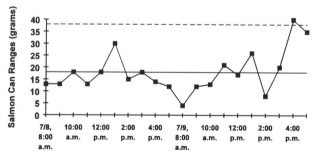

- **Interpret the results**

    Both the $\overline{X}$ chart and the R chart appear to be in control with no apparent trends or patterns until the last two subgroups. Also measurements taken indicate the minimum specification of 227 grams is being met. However the data from July 9th at 4:00 p.m. and 5:00 p.m. signal an out of control process. At 4:00 p.m., the range was 40 which was higher than the UCL_R, and at 5:00 p.m., the mean weight was 252.2 which was higher than the $UCL_{\overline{X}}$.

    Action taken: if this control chart had already been set up, immediate action would have been taken at 4:00 p.m. on July 9th to try to solve the problem. However the chart was not set up until after the 5:00 p.m. reading. This is a summary of the action taken after the 5:00 p.m. reading.

    It was decided to weigh samples of cans from each of the filler machine's two heads. Five cans were taken from each head with the following results:

	1	2	3	4	5	Mean
Head 1	235	238	230	234	240	**235.4**
Head 2	263	263	269	270	265	**266.0**

    Obviously, Head 2 was causing the problem making the range and mean go out of control. Head 2 was adjusted to match more closely the weight results from Head 1.

    The production line continued operation the next day. Samples were taken every 15 minutes during the first hour of operation. After it was determined the process was back in control, the readings went back to being taken every hour.

2.4. Interpreting Control Charts

A main purpose of using control charts is to identify when process changes have occurred. As shown in the example, this is signaled by having a point outside the control limits. However there are other signals to look for when analyzing control charts. These signals are summarized in Figure 18.3.

Figure 18.3. Various Signals which Identify Process Change

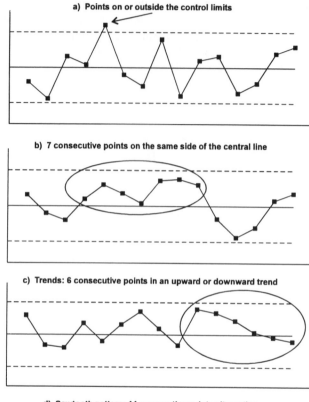

a) Points on or outside the control limits

b) 7 consecutive points on the same side of the central line

c) Trends: 6 consecutive points in an upward or downward trend

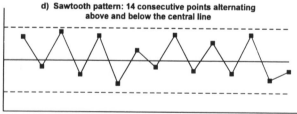

d) Sawtooth pattern: 14 consecutive points alternating above and below the central line

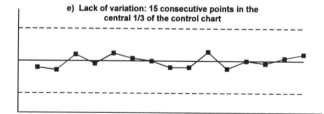

e) Lack of variation: 15 consecutive points in the central 1/3 of the control chart

2.5. Actions To Take

As a general rule of thumb, if a process change has occurred according to the control charts, it is prudent to identify an assignable cause. Your goal as an engineer is to eliminate undesirable assignable causes to bring the process under control. However this requires intimate technical knowledge of the process. The control chart tells only when to look for an assignable cause but not where to look or what cause will be found.

2.5.1. Changes in Sample Means Only

Take the case where the $\overline{X}$ control chart signals a process change but the R control chart does not. Some possible areas to begin your search are:

- change in materials or subassemblies coming from different sources
- gradual deterioration of processing equipment
- deterioration or changes in environmental conditions such as temperature
- accumulation of waste products
- worker fatigue.

Actions to reduce process shifts often call for fundamental changes in equipment or methods used in the process.

2.5.2. Changes in Sample Ranges Only

Take the case where the R control chart signals a process change but the $\overline{X}$ control chart does not. This case is particularly likely to be found in processes where operator skill and care are an important factor. Start your investigation by examining how various operators perform the particular process.

2.5.3. Changes in Sample Means and Ranges

Take the case where both the $\overline{X}$ and R control charts signal a process change. This is common in the first stages of control chart use. The important point is not to get discouraged. The control chart results should be viewed as an indication that further improvement is possible.

Begin your investigation by looking for simple causes. Check to make sure the control limit calculations were done correctly. Then check to see if the variations were due to measurement errors.

If these simple checks do not uncover why the control charts signal a process change, then a more detailed investigation is in order. As stated previously, detailed investigations looking for assignable causes require intimate technical knowledge of the process. Design of Experiments is a tool that can be used to help in these investigations. DOE will be described in the next chapter.

2.6. Continued Use of the Control Charts

Once a control chart has been established, it is important to continue its use. Plot new subgroup results immediately after measuring so any assignable cause can be identified as quickly as possible. Continued use of the control charts also means periodic review of the central lines and control limits.

2.6.1. Case 1: Process In Control

If the process is in control, the trial control limits that were established based on the first 20 subgroups can be applied to future production. A regular period should be established to review and possibly recalculate the control limits. The regular period for review can be based on time, such as every week or month, or it can be based on the number of subgroups, such as every 20, 50, or 100 subgroups. A recalculation of the control limits should be performed if a process change has occurred.

2.6.2. Case 2: Process Out of Control

For the case where the $\overline{X}$ chart and/or the R chart shows the process is out of control, the engineer has two alternatives:

1) Continue to use the trial control limits for future production.

2) Recalculate the central line and control limits by eliminating the out of control points.

The choice made by the engineer is somewhat subjective based on the situation, but a rule of thumb is continue to use the same control limits if an assignable cause has not been identified and eliminated. If the assignable cause has been identified and eliminated, recalculate the central line and control limits by not counting the out of control points.

Also the control limits should be recalculated for any intentional process change which results in an $\overline{X}$ or R change. Once the change has occurred, gather data from the next 20 subgroups and recalculate.

A regular period based on time or number of subgroups should be established to review the control limits. If the process is out of control, this review should be more frequent than if the process is in control. The out of control process requires more attention, maintenance, and effort.

3. $\overline{X}$, S CONTROL CHARTS

$\overline{X}$, S control charts are used when the data are variable data, and the sample size of the subgroups is 15 or greater.

The $\overline{X}$, S control charts are almost identical to the $\overline{X}$, R control charts. The same outline shown in Table 18.1 is used to construct them. The objective and interpretation of the control charts, actions to take, and continued use of the control charts are all the same. The difference is that an S chart is used rather than an R chart to estimate the noise or common cause variation of the $\overline{X}$ chart. The sample size of the subgroups for an S chart is 15 or greater whereas the sample size of the subgroups for an R chart is less than 15. When the sample size is 15 or greater, the sample size is large enough so that the sample standard deviation provides a better estimate of the noise compared to using the sample range.

3.1. Subgroup Calculations

After the data has been taken from about 20 subgroups, there are two statistics to calculate from each subgroup: the sample mean ($\overline{X}$) and the sample standard deviation (S). The sample mean was shown previously.

$$\overline{X} = \frac{\sum_{i=1}^{N} X_i}{N} \tag{16.1}$$

The sample standard deviation is the square root of the sample variance.

$$S = \sqrt{\frac{\sum_{i=1}^{N} \left(X_i - \overline{X} \right)^2}{N - 1}} \tag{18.8}$$

where N is the number of elements per subgroup.

3.2. Control Limit Calculations

To calculate the control limits, take the sample means and sample standard deviations that were calculated for each subgroup and compute the mean of the sample means ($\overline{\overline{X}}$) and the mean of the sample standard deviations ($\overline{S}$).

$$\overline{\overline{X}} = \frac{\sum \overline{X}}{n} \tag{18.2}$$

$$\overline{S} = \frac{\sum S}{n} \tag{18.9}$$

where n is the number of subgroups.

Then use Table 18.3 to calculate the upper and lower control limits for the $\overline{X}$ and S charts. The control limits for the $\overline{X}$, S charts represent the 3 sigma variation limits for the distribution of $\overline{X}$ values and S values, respectively.

3.3. $\overline{X}$, S Control Chart Example

During the process of making an integrated circuit, one of the steps involves depositing a thin insulator layer. The target thickness is 100 nm, and it is important to minimize variations. If the layer is too thin, it will be too weak to withstand the voltage applied across it. If it is too thick, a subsequent etching step will not etch all the way through the layer.

- **Objective**

    The primary objective is to quickly identify any data points that are out of control in order to minimize variations.

Table 18.3._Control Limit Calculations for $\overline{X}$, S Control Charts

$\overline{X}$ chart	S chart
Central line = $\overline{\overline{X}}$	Central line = $\overline{S}$
$UCL_{\overline{X}} = \overline{\overline{X}} + A_1\overline{S}$	$UCL_S = B_4\overline{S}$
$LCL_{\overline{X}} = \overline{\overline{X}} - A_1\overline{S}$	$LCL_S = B_3\overline{S}$

Number of observations in subgroup	A_1	B_3	B_4
15	0.82	0.43	1.57
16	0.79	0.45	1.55
17	0.76	0.47	1.53
18	0.74	0.48	1.52
19	0.72	0.50	1.50
20	0.70	0.51	1.49
21	0.68	0.52	1.48
22	0.66	0.53	1.47
23	0.65	0.54	1.46
24	0.63	0.55	1.45
25	0.62	0.56	1.44
30	0.56	0.60	1.40
35	0.52	0.63	1.37
40	0.48	0.66	1.34
45	0.45	0.68	1.32
50	0.43	0.70	1.30
55	0.41	0.71	1.29
60	0.39	0.72	1.28
65	0.38	0.73	1.27
70	0.36	0.74	1.26
75	0.35	0.75	1.25
80	0.34	0.76	1.24
85	0.33	0.77	1.23
90	0.32	0.77	1.23
95	0.31	0.78	1.22
100	0.30	0.79	1.21

- **Determine subgroupings**

    The insulator is deposited on one lot at a time. Each lot consists of 25 wafers. There is only one machine to do this deposition process in production.

    To get the best indication of the results of any deposition, measurements need to be taken at various locations on the wafers. Also at least 5 wafers should be measured from each lot, taken at random.

    Conclusion: the subgroupings will consist of each wafer lot. The insulator thickness of 5 wafers will be measured at the top, center, and bottom locations of each wafer. Thus each subgroup will consist of 15 measurements, and an $\overline{X}$, S chart will be used.

- **Collect measurement data and calculate** $\overline{X}$, S **for each subgroup**

Lot	$\overline{X}$	S
1	101.0	6.90
2	102.3	5.23
3	98.7	4.14
4	98.0	2.85
5	102.9	4.81
6	103.8	5.49
7	101.4	6.10
8	99.2	3.84
9	99.0	4.84
10	100.0	5.86
11	100.9	5.55
12	98.6	3.90
13	102.8	6.21
14	101.8	4.36
15	114.5	5.99
16	102.2	4.02
17	115.9	4.01
18	99.5	3.85
19	98.3	6.05
20	100.2	6.00
Mean	**102.1**	**5.00**

- **Calculate control limits**

    $\overline{\overline{X}} = 102.1$

    $\overline{S} = 5.00$

$$UCL_{\overline{X}} = \overline{\overline{X}} + A_1 \overline{S}$$

$A_1 = 0.82$ for subgroup size of 15 (from Table 18.3)

$$UCL_{\overline{X}} = 102.1 + (0.82)(5.00) = 106.2$$

$$LCL_{\overline{X}} = \overline{\overline{X}} - A_1 \overline{S} = 102.1 - (0.82)(5.00) = 98.0$$

$$UCL_S = B_4 \overline{S}$$

$B_4 = 1.57$ for subgroup size of 15 (from Table 18.3)

$$UCL_S = (1.57)(5.00) = 7.85$$

$$LCL_S = B_3 \overline{S}$$

$B_3 = 0.43$ for subgroup size of 15 (from Table 18.3)

$$LCL_S = (0.43)(5.00) = 2.15$$

- **Plot the data**

$\overline{X}$ Chart: Insulator Thicknesses

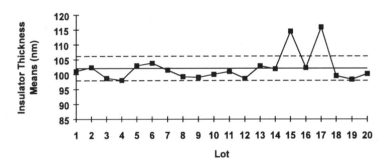

S Chart: Insulator Thicknesses

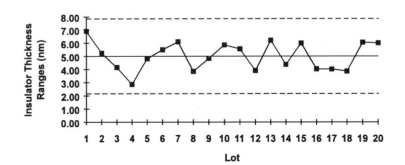

* **Interpret the results**

There are two data points outside the control limits—the $\overline{X}$ values for lots 15 and 17. The $\overline{X}$ values for these lots were 114.5 and 115.9, respectively, compared with an upper control limit of 106.2. Now that these problem lots have been identified, the next step is to identify the assignable causes and solve the problem. This will be done with the help of Design of Experiment tools in the next chapter. This case is covered in the section on Resolution IV matrix designs (Chapter 19, Section 4.2.1.3.).

The other action to take once the assignable cause is identified and eliminated is to recalculate the control limits without counting the data from lots 15 and 17. This is for the purpose of establishing control limits that can be used when plotting data from lots processed in the future.

$$\overline{\overline{X}} = 100.6$$

$$\overline{S} = 5.00$$

$$UCL_{\overline{X}} = \overline{\overline{X}} + A_1\overline{S} = 100.6 + (0.82)(5.00) = 104.7$$

$$LCL_{\overline{X}} = \overline{\overline{X}} - A_1\overline{S} = 100.6 - (0.82)(5.00) = 96.49$$

$$UCL_S = B_4\overline{S} = (1.57)(5.00) = 7.85$$

$$LCL_S = B_3\overline{S} = (0.43)(5.00) = 2.15$$

4. PROCESS CAPABILITY

Process capability is measured by comparing the process data with the upper and lower specification limits for the process. To the extent the process data fall within the specification limits, the process is deemed capable. Process capability can be a function of both the spread and centering of the process data. This is illustrated in Figure 18.4. First consider Distribution A. It has a narrow spread and is centered. Distribution A falls well within the specification limits and the process is considered capable. Now consider Distribution B which has the same amount of spread as Distribution A, but it is not centered. The process with Distribution B is also considered capable because it falls within the specification limits. Thus when the amount of spread is small relative to the specification limits, the process does not need to be centered to be capable.

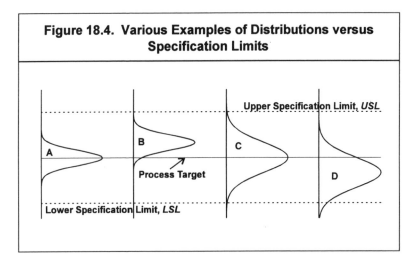

Figure 18.4. Various Examples of Distributions versus Specification Limits

Now consider Distribution C. Its spread is much wider than Distributions A and B. The process with Distribution C is also capable, but there is less room for error. Distribution C must be centered on the process target in order to be capable. Finally consider Distribution D. It has the same spread as Distribution C, but it is not centered. The process represented by Distribution D is not capable because part of the distribution falls outside the specification limits.

4.1. Two-Sided Specifications Considering Only Spread

One measure of process capability is the C_P index. The C_P index measures the process spread relative to the specification limits.

$$C_P = \frac{USL - LSL}{6\sigma} \tag{18.10}$$

where *USL* is the upper specification limit, *LSL* is the lower specification limit, and σ is the standard deviation of the population. The 6σ term in the denominator is used because the C_P index compares a two-sided spread of the data to a two-sided specification. Table 18.4 predicts what the fallout will be for various values of C_P assuming the process is centered. A bare minimum requirement for process capability is to have a C_P of at least 1.0. However this still leads to 2700 ppm defective. A more ideal manufacturing situation is to have a C_P of at least 1.3. This brings the predicted defective rate below 100 ppm.

Table 18.4. Prediction of Process Data Falling Outside Specification Limits vs. C_P for a Centered Process

C_P	Parts per million defective
0.50	133,600
0.75	24,400
1.00	2,700
1.10	967
1.20	318
1.30	96
1.40	26
1.50	6.8
1.60	1.6
1.70	0.34
1.80	0.06
2.00	0.0018

One of the limitations of the C_P index is that it does not take into consideration the centering of the process. To account for the centering as well as the spread of the distribution, let's first define two terms that are used to measure the process capability for single-sided specifications.

4.2. Single-Sided Specifications

C_{PL} is the process capability index for a specification that defines a lower limit. An example is a voltage specification that calls for it to be 4.0 V or greater.

$$C_{PL} = \frac{\overline{\overline{X}} - LSL}{3\sigma} \tag{18.11}$$

C_{PU} is the process capability index for a specification that defines an upper limit. An example is a voltage specification that calls for it to be 6.0 V or lower.

$$C_{PU} = \frac{USL - \overline{\overline{X}}}{3\sigma} \tag{18.12}$$

The 3σ term is used in the denominator for C_{PL} and C_{PU} because these indices compare a one-sided spread of the data to a one-sided specification.

4.3. Two-Sided Specifications Considering Spread and Centering

Now to account for spread and centering with a two-sided specification, the term C_{PK} is used.

$$C_{PK} = \text{the lower of two values, } C_{PL} \text{ or } C_{PU} \qquad (18.13)$$

C_{PK} is considered a more severe measure than C_P for measuring process capability because it considers centering as well as the spread of the distribution. Again as with C_P, the bare minimum requirement to deem a process capable is for C_{PK} to be at least 1.0. However a more ideal situation is for C_{PK} to be at least 1.3.

A final point to keep in mind is that upper and lower specification limits are not the same as upper and lower control limits. The specification limits are used to judge whether or not the process is capable, whereas the control limits are used to judge whether or not the process is in control. It is possible for a process to be capable but out of control, and it is also possible to be not capable but in control. A goal of engineers is to develop and maintain processes that are capable and in control.

4.4. Process Capability Example

This example is a continuation of the salmon cannery example. By using $\overline{X}$, R control charts, it was found that the last two subgroupings were out of control. In this example, the process capability of the salmon filling process will be calculated. Recall the target weight for each can's contents is 242 grams. The upper specification limit (*USL*) is 257 grams, and the lower specification limit (*LSL*) is 227 grams.

- **Objective**

  To determine the process capability of the salmon can filling process.

- **Collect measurement data**

  The same data that were collected to construct the $\overline{X}$, R chart will be used. This was shown in the $\overline{X}$, R chart example. Each subgroup consisted of five cans selected at random. Samples were taken once an hour. The data of 20 subgroups were gathered for a total of 100 data points. Recall the process was found to be out of control for the last two subgroups. The reason for being out of control was traced to Head 2 of the filler machine, and this was corrected. Since the cause for the process to be out of control was identified and corrected, only the data taken while the process was in control will be used for the calculation. These data are from the first 18 subgroups.

Time	1	2	3	4	5	$\overline{X}$	S^2
7/8, 8:00 a.m.	238	243	235	230	235	236.2	22.7
9:00 a.m.	242	235	229	240	241	237.4	29.3
10:00 a.m.	250	239	252	241	234	243.2	57.7
11:00 a.m.	235	234	247	245	241	240.4	33.8
12:00 p.m.	231	246	243	228	235	236.6	59.3
1:00 p.m.	260	246	230	240	240	243.2	121.2
2:00 p.m.	237	230	245	233	243	237.6	40.8
3:00 p.m.	229	245	247	240	233	238.8	59.2
4:00 p.m.	234	238	239	230	244	237.0	28.0
5:00 p.m.	250	245	243	241	238	243.4	20.3
7/9, 8:00 a.m.	243	239	239	241	243	241.0	4.0
9:00 a.m.	243	243	231	236	238	238.2	25.7
10:00 a.m.	246	247	234	236	244	241.4	35.8
11:00 a.m.	246	235	230	236	251	239.6	74.3
12:00 p.m.	243	247	235	234	230	237.8	48.7
1:00 p.m.	238	230	231	231	256	237.2	120.7
2:00 p.m.	238	239	242	246	243	241.6	10.3
3:00 p.m.	247	235	235	255	235	241.4	84.8
					Mean	239.6	48.7

The overall average ($\overline{\overline{X}}$) is 239.6.

Each subgroup has an estimate of variance with 4 degrees of freedom. To calculate the overall variance, pool the estimates of variance by taking the weighted average using the following formula.

$$S_{avg}^2 = \frac{\sum S_i^2 \phi_i}{\sum \phi_i}$$

The overall variance (S_{avg}^2) is 48.7. This estimate consists of $4 \times 18 = 72$ degrees of freedom. Since $\phi > 60$, the population variance (σ^2) is known for all practical purposes and is equal to 48.7.

- **Interpret the results**

    These are the known values for this section.

    $\overline{\overline{X}} = 239.6$ grams

    $\sigma^2 = 48.7$ $\quad\quad\quad\quad\quad$ $\sigma = \sqrt{48.7} = 6.98$ grams

    $USL = 257$ grams

    $LSL = 227$ grams

    First, calculate the process capability index.

    $$C_P = \frac{USL - LSL}{6\sigma} = \frac{257 - 227}{6(6.98)} = 0.72$$

    However, the process is not quite centered. The target is 242, and the overall average weight is slightly lower at 239.6. Thus to get the worst case measure of process capability, C_{PK} will be calculated.

    C_{PK} = the lower of two values, C_{PL} or C_{PU}

    $$C_{PL} = \frac{\overline{\overline{X}} - LSL}{3\sigma} = \frac{239.6 - 227}{3(6.98)} = 0.60$$

    $$C_{PU} = \frac{USL - \overline{\overline{X}}}{3\sigma} = \frac{257 - 239.6}{3(6.98)} = 0.83$$

    Therefore, $C_{PK} = 0.60$

    This is not an acceptable manufacturing process. The ideal situation is to have a process capability index of 1.3 or greater.

    Thus the salmon can filling process was out of control and not capable. From the earlier example, the process was later brought back into control by adjusting one of the heads of the filling machine. However, the process is still not capable. With the natural variation of the process and the current specification limits, there will be product that falls outside the specification limits. There is a mismatch between the specification limits and the process spread.

    There are three alternatives facing the engineer in this situation.

1) **Attempt to decrease the process variation.**
Utilize Design of Experiment techniques to quantify which variables of the filling process lead to the largest process variation of salmon can weights. This will provide clues as to what parameters need to be controlled more tightly.

2) **Examine the upper and lower specification limits to see if they can be modified to provide a wider process window.**
In this case, the lower specification limit is fixed because it is printed on the can label, but perhaps the specification on the label can be lowered. Also, the upper specification was set to ensure the proper vacuum could be obtained and to minimize waste. However, a close examination could be conducted to verify whether or not a satisfactory vacuum could be obtained for weights greater than 257 grams.

3) **Make no changes, and live with the risks of producing product outside the specification limits.**
This is an alternative in the real world. Management may be willing to take this risk. At least by calculating process capability indices, the risks can be quantified, and the various manufacturing processes can be compared to pinpoint the most critical process steps.

5. *p* CONTROL CHARTS

p control charts are used when the data are attribute data. In practice, one of the limitations of control charts for variables is that it is often unreasonable to have a control chart for each quality characteristic in a manufacturing process. This could lead to hundreds of $\overline{X}$, R or $\overline{X}$, S charts. Another limitation is that some quality characteristics cannot be expressed in terms of numbers; they can be observed only as attributes.

Thus there is a need for an attributes control chart, and the most widely used is the *p* chart. This is a chart for the fraction rejected as nonconforming to specifications.

As with the $\overline{X}$, R and $\overline{X}$, S control charts, Table 18.1 can be used as an outline to construct p charts. However there are two key distinctions when working with an attributes control chart versus a variables control chart. One distinction is the population is based on a binomial distribution. Another distinction is that subgroups consist of entire lots of material.

5.1. Subgroup Calculations

To repeat, the subgroup for a p chart consists of an entire lot of material. Examples of lots are the production from each day or the production made from a particular batch of subassemblies. Once the subgroup goes through the inspection or testing step desired for this control chart, the fraction rejected (p) is calculated.

$$p = \frac{\text{Number of rejects}}{N} \tag{18.14}$$

where N is the number of test results in the subgroup.

5.2. Control Limit Calculations

The first step to calculate the control limits is to sum the total number of rejects and the total number tested for all the subgroups. Then calculate the mean fraction rejected ($\overline{p}$).

$$\overline{p} = \frac{\text{Total number of rejects}}{\text{Total number tested}} \tag{18.15}$$

The p chart is based on a binomial distribution. However because of the phenomenon explained by the central limit theorem, the distribution of the fraction rejected from various subgroups will be normally distributed around a central value even though the distribution of the population is binomial. The standard deviation of a binomial distribution is $\sqrt{\overline{p}(1 - \overline{p})} \, / \, N$. Thus the control limits for the p chart are as follows.

$$UCL_p = \overline{p} + 3\sigma_p = \overline{p} + 3\sqrt{\frac{\overline{p}(1 - \overline{p})}{N}} \tag{18.16}$$

$$LCL_p = \bar{p} - 3\sigma_p = \bar{p} - 3\sqrt{\frac{\bar{p}(1 - \bar{p})}{N}} \tag{18.17}$$

where N is the number tested in each subgroup. Since it is possible the number tested in each subgroup may vary because of different lot sizes, it is likely the control limits will vary from subgroup to subgroup.

5.3. *p* Control Chart Example

The manufacturing of charge-coupled devices (CCDs) to make imaging systems involves many processes: processing silicon wafers, sawing the wafers, making electrical connections, etc. Each of these processes have performance characteristics that could be tracked, but it would take many control charts to do so. It is decided for the sake of simplicity to use a *p* chart to monitor the amount of rejects after the imaging system is assembled. This final test involves many tests such as brightness, contrast, and detection of defects on the test screen. The imaging system can be rejected for failing to meet any one of these tests. All systems are subjected to final testing.

- **Objective**

    There are two objectives for setting up the control chart:

    1) To discover the average fraction of imaging systems rejected.

    2) To discover points that exceed the upper control limit for the fraction rejected and call for corrective action.

- **Determine subgroupings**

    There are imaging systems which finish processing every day. However the number of systems completed depends on demand and varies widely from day to day.

    Conclusion: the subgroupings will consist of the imaging systems completed each day. All systems will go through final testing. Each system that fails to pass any one of the final tests will be rejected. The failure categories will be noted for later use.

- **Collect measurement data and calculate p for each subgroup**

  The following table shows the results from the first 18 subgroups.

Day	Systems Tested	Rejects	Fraction rejected, p	UCL_p	LCL_p
1	179	4	2.23%	9.60%	0.01%
2	884	58	6.52%	6.96%	2.65%
3	868	37	4.28%	6.98%	2.63%
4	611	37	5.97%	7.40%	2.21%
5	417	28	6.71%	7.95%	1.66%
6	307	21	6.81%	8.47%	1.14%
7	442	24	5.43%	7.86%	1.75%
8	228	10	4.49%	9.05%	0.56%
9	618	42	6.86%	7.39%	2.22%
10	667	37	5.54%	7.29%	2.32%
11	589	35	5.92%	7.45%	2.16%
12	688	16	2.33%	7.25%	2.36%
13	736	43	5.82%	7.17%	2.44%
14	525	12	2.34%	7.61%	2.00%
15	269	7	2.75%	8.72%	0.89%
16	534	17	3.22%	7.58%	2.03%
17	608	20	3.30%	7.41%	2.20%
18	310	7	2.27%	8.45%	1.16%
Totals	**9480**	**456**	**4.81%**		

- **Calculate control limits**

$$\overline{p} = \frac{\text{Total number of rejects}}{\text{Total number tested}} = \frac{456}{9480} = 0.0481$$

$$UCL_p = \overline{p} + 3\sigma_p = \overline{p} + 3\sqrt{\frac{\overline{p}(1 - \overline{p})}{N}}$$

$$UCL_p = 0.0481 + 3\sqrt{\frac{(0.0481)(1 - 0.0481)}{N}} = 0.0481 + \frac{0.6419}{\sqrt{N}}$$

where N is the number in each subgroup.

$$LCL_p = \overline{p} - 3\sigma_p = 0.0481 - \frac{0.6419}{\sqrt{N}}$$

The upper and lower control limits are calculated for each subgroup and are shown in the preceding table.

- **Plot the data**

p Chart: Imaging System Rejects

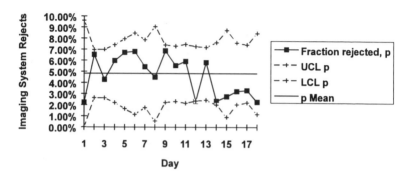

- **Interpret the results**

The first item to note is the upper and lower control limits are different for each subgroup because the number of systems tested are different for each subgroup.

The second item to note is the fraction rejected is in control. The fraction rejected for each subgroup falls within the control limits. During the first 11 days, the fraction rejected was greater than p for most of the days. However there were not 7 days in a row above p which would have signaled an out of control process.

Also the last 5 days have been below $\bar{p}$. The fraction rejected for the next 2 days should be watched closely to see if they are also below p. If they are, this would make 7 consecutive days below p, and this would signal an out of control process. Keep in mind an out of control process with the fraction rejected below p for 7 consecutive subgroupings is not a negative event. However this would be an indication that an assignable cause should be sought in order to maintain the process at an improved level.

Conclusion: the process is in control for the first 18 days. The same p will be used to calculate control limits for future subgroupings.

6. PARETO ANALYSIS

Pareto analysis is a graphical method to show the highest to lowest significance of events or occurrences. It is a valuable tool that tells where to focus your efforts when solving a problem. The general rule of thumb governing Pareto analysis is 80% of the problems come from 20% of the causes. Therefore when solving problems, concentrate your effort on the prominent causes. Pareto analysis can be used in conjunction with p charts to help find causes of rejects.

To construct a chart for Pareto analysis, some more information is required from each of the subgroups. Along with keeping track of the number of rejects, also categorize the reason for each reject. The reject categories chosen depends on the types of rejects seen and requires a technical understanding of the test. After a number of subgroups have been tested, sum up the total number of rejects by category. Rank the reject categories from the highest to lowest number of occurrences, and plot in a histogram. For added information, you can plot a cumulative percentage of the reject categories.

6.1. Pareto Analysis Example

This example expands upon the situation described in the p chart example. Recall this example involves the production of imaging systems. The subgroups were divided by days of production. For each subgroup, the number of rejects were counted as well as the number of rejects in each category.

- **Objective**

  The first objective is to identify the most common reject categories. This will provide clues where to focus efforts when trying to decrease the number of rejects.

  A second objective is to identify the reject signature for the process while it is in control. This serves as a comparison point for when the process goes out of control. If and when the process goes out of control, the reject signature of the out of control process can be compared with the reject signature of the controlled process in order to provide clues as to what changes have occurred.

- **Collect measurement data**

  The table below shows the results from the first 18 subgroups and the totals by reject category.

Day	Systems Tested	Rejects	Point Defects	Column Defects	Brightness	Contrast	Other
1	179	4	2	1	1		
2	884	58	30	14	7	4	3
3	868	37	20	15		2	
4	611	37	12	13	8	2	2
5	417	28	21	7			
6	307	21	12	5	3		1
7	442	24	10	8	4	2	
8	228	10	8	2			
9	618	42	25	8	5	4	
10	667	37	16	10	4	4	3
11	589	35	19	12	4		
12	688	16	7	5	4		
13	736	43	20	9	5	5	4
14	525	12	7	5			
15	269	7	5	2			
16	534	17	8	5	4		
17	608	20	10	7	3		
18	310	7	5	2			
Totals	9480	456	237	130	52	23	13

- **Plot the data**

Pareto Chart: Imaging System Reject Categories

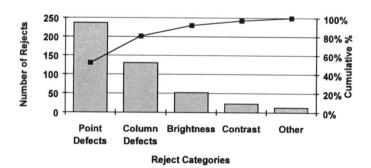

This figure shows the total number of rejects by category as well as the cumulative % of the reject categories.

- **Interpret the results**

    The most common reject category is point defects in the imaging system accounting for 52% of the rejects. Reducing the number of point defect rejects is obviously the place to begin in order to decrease the fraction of imaging systems rejected.

    The next most common reject category is column defects in the imaging system at 29% of the rejects. The point and column defects combined account for 81% of the rejects. The rest of the reject categories are not nearly as prominent. Rejects for brightness, contrast, and for other causes account for 11%, 5%, and 3% of the rejects, respectively.

    Now that this reject signature has been determined, it is possible to focus on the most important problem, in this case point defects, in order to decrease the overall reject rate. It is also possible to compare the reject signature of any out of control subgroups to this reject signature to give information as to why the fraction rejected went out of control.

    This concludes the chapter on SPC. This chapter has shown how you can utilize SPC to monitor and control production. The main tools are various types of control charts: $\overline{X}$, R and $\overline{X}$, S control charts for variables data and p control charts for attributes data. The primary purpose of the control charts is to separate signals from the noise. A signal indicates there exists an assignable cause, and you should take action to identify it. In addition, the concepts of process capability and Pareto analysis were explained. These concepts are additional tools that can help you analyze and identify problems in a production process. These tools are summarized in Table 18.5.

Table 18.5. Summary of Statistical Process Control Tools

Item	Type of Data	Purpose
$\overline{X}$, R control charts	Variable data Subgroup sample size < 15	To identify assignable causes of variation
$\overline{X}$, S control charts	Variable data Subgroup sample size ≥ 15	To identify assignable causes of variation
p control chart	Attribute data	To provide a basis for acceptance or rejection of manufactured product and to identify assignable causes of variation
C_P and C_{PK} indices	Specification limits and process data	To analyze a process to see if certain specifications can be met
Pareto chart	Number of occurrences in categories	To identify where to focus your problem solving efforts

Chapter 19

DESIGN OF EXPERIMENTS

Design of Experiments is a key tool for engineers to solve problems. However one of the difficulties with practical application of DOE is experiments are usually planned poorly. The first section on DOE explains how an experiment should be planned. The second section describes the steps for designing an experiment. The third and fourth sections cover designs for single variable experiments and multi-variable experiments, respectively.

1. EXPERIMENT STRATEGY

Engineers typically receive little training in devising strategies for experiments. Incorrect strategies can lead to incorrect decisions made about the experiment. This can snowball into finger pointing sessions: manufacturing engineering blaming their problems on development engineering and vice versa.

However with sound utilization of DOE principles, misinterpretation of the data will be minimized, and correct technical decisions will be made. The trick is to define the objective of the experiment as specifically as possible. In research, the objective may be to understand a process that has never been studied before. In development, the objective may be to see if a product can meet a certain specification, and this specification needs to be quantified. In manufacturing, the objective may be to find the cause of an out of control process or to see if a certain cost reduction idea can be employed. To repeat, this objective needs to be stated as specifically as possible.

Once the objective is defined, it is important to spend time defining the experimental space in writing. The experimental space is the amount variables affect the result as they are changed. This definition of experimental space should be based on the engineer's technical knowledge of the product or process. It should specify all possible variables that may have an impact on the response of interest. The definition should also specify other responses that may be affected by changes in these variables.

The experiment strategy differs depending on whether the experiment is in research and development or manufacturing engineering. In research and development, the experimental space may be largely unknown. The experiment strategy may consist of two experiments. The first experiment identifies a starting point and helps identify the most significant variables. Then a second experiment can be performed to find out more about the experimental space such as identifying interactions. In manufacturing, the experimental space is likely to be more defined based on past experience. Thus the experiment strategy for a manufacturing experiment would likely consist of one experiment which identifies the most significant variables and interactions.

Regardless of whether the experiment is part of a research and development or a manufacturing effort, the purpose of the experiment is to reach a decision with a known confidence level. With the correct experiment strategy, the decision should be clear cut and result in immediate action. For instance a common question answered by DOE is whether or not it is feasible to introduce a new product or process into manufacturing. If it is determined by DOE to be not feasible, the product or process should be killed immediately to minimize the cost and time spent on it. On the other hand, if it is determined to be feasible, the project should be completed as soon as possible in order to reap the maximum benefit for the minimum cost.

2. STEPS FOR DESIGNING AN EXPERIMENT

The steps for designing an experiment are outlined in Table 19.1.

2.1. State the Objective

The first step is to state the objective. During an experiment, the experimenter usually selects test samples, uses statistics to analyze test

Table 19.1. Outline for Designing an Experiment

1. Define the objective by stating 2 alternative decisions:

    H_0 (null hypothesis)

    H_a (alternative hypothesis)

2. Define the acceptable risk for selecting the wrong alternative:

    α error (accepting H_a when H_0 is true)

    β error (accepting H_0 when H_a is true)

3. Determine the significant difference sought: δ

4. Determine the variables to be included in the experiment

5. Calculate the appropriate sample size and determine how each sample will be processed

6. Calculate the test criterion for selecting between the alternative decisions. If variances are unknown or cannot be estimated prior to the experiment, do this step after performing the experiment.

7. Perform the experiment and calculate the statistics

8. Compare experimental data with the test criterion

9. Make the decision

results, and tries to make a comparison between two populations to see if they are the same or different. Defining the objective is done by stating two alternatives, and the objective of the experiment is to decide between these two alternatives. These alternatives are the null hypothesis (H_0) and the alternative hypothesis (H_a).

Null hypothesis (H_0): there is no significant difference between the two populations for the parameter of interest.

Alternative hypothesis (H_a): there is a significant difference between the two populations for the parameter of interest.

This chapter concentrates on experiments where decisions are made about population means. The null hypothesis can be stated in terms of one population compared to another or one population compared to a fixed value (μ_0) such as a specification. Possible null hypotheses are as follows.

H_0: $\mu_1 = \mu_2$ Comparing one population with another population.

H_0: $\mu_1 = \mu_0$ Comparing one population with a fixed value.

The alternative hypothesis can also be stated in terms of one population compared to another or in terms of one population compared to a fixed value. An added twist is the alternative hypothesis can also be stated as two populations not equal (double-sided hypothesis) or as one population mean greater than or less than the other population (single-sided hypothesis). Thus there are several ways an alternative hypothesis can be stated, and it is up to you how to state it depending on what you want to prove. Possible alternative hypotheses are as follows.

H_a: $\mu_1 \neq \mu_2$ Comparing one population with another population (double-sided).

H_a: $\mu_1 \neq \mu_0$ Comparing one population with a fixed value (double-sided).

H_a: $\mu_1 > \mu_2$ Comparing one population with another population (single-sided).

H_a: $\mu_1 < \mu_2$ Comparing one population with another population (single-sided).

H_a: $\mu_1 > \mu_0$ Comparing one population with a fixed value (single-sided).

H_a: $\mu_1 < \mu_0$ Comparing one population with a fixed value (single-sided).

2.2. Define the Risks

Once the objective has been stated in terms of null and alternative hypotheses, the next step is to define the risks. The experimenter runs the risk of making one of two errors in any experiment: an alpha error (α) or a beta error (β).

Alpha error (α): the error of accepting the alternative hypothesis (H_a) when the null hypothesis (H_0) is actually true.

Beta error (β): the error of accepting the null hypothesis (H_0) when the alternative hypothesis (H_a) is actually true.

To help clarify these errors, consider the probability matrix shown in Table 19.2 that is applicable before an experiment is performed. Suppose for example an experiment is set up to see if the population mean (μ_1) can be significantly increased compared to a fixed value (μ_0). The experimenter sets up the experiment with $\alpha = 5\%$ and $\beta = 10\%$. First consider the case where the true state of nature is no increase in μ_1. There is a 95% probability ($1 - \alpha$) the correct decision will be made to accept the null hypothesis. There is a 5% probability (α) the wrong decision will be made by accepting the alternative hypothesis.

Second consider the case where the true state of nature is an increase in μ_1. There is a 90% probability ($1 - \beta$) the correct decision will be made to accept the alternative hypothesis, and there is a 10% probability (β) the wrong decision will be made by accepting the null hypothesis.

It is up to the experimenter to decide the α and β errors when setting up the experiment. This decision should be based on practical issues. With smaller values of α and β, samples sizes increase. It is often impractical to choose α and β errors of 0% because this means entire populations need to be tested. It also does not make sense to choose α and β errors of 50%. The experimenter would do just as well to flip a coin rather than going through the trouble of setting up the experiment.

Table 19.2. Probability Matrix for an Experiment Before It Is Conducted (Single-Sided Alternative Hypothesis)

		Experimenter's decision	
		Accept H_0: $\mu_1 = \mu_0$	Accept H_a: $\mu_1 > \mu_0$
True state of nature	$\mu_1 = \mu_0$	Correct decision Probability = $(1-\alpha)$	An α error Probability = α
	$\mu_1 > \mu_0$	A β error Probability = β	Correct decision Probability = $(1-\beta)$

Therefore the values of α and β selected need to be between 0% and 50%. The decision of what values to select depend on the situation. The risks of making the wrong decision need to be weighed against the costs of performing the experiment.

2.3. Define the Significant Difference

The experimenter's next step is to define the significant difference sought. This difference is designated as δ. For instance, suppose an engineer wants to improve the bond strength of a wire bonding operation. The mean bond strength is μ_1. The engineer considers if the bond strength can be improved an amount δ by increasing the bonding temperature, the change would be cost effective. Therefore the experiment is designed to decide whether or not an increase in bonding temperature will increase the bond strength to $(\mu_1 + \delta)$ or greater.

As with the choice of α and β values, selecting a δ value depends on the situation. The number of required samples increases as δ decreases.

2.4. Determine the Variables

Determining the variables to include in the experiment is part of the experiment strategy. The number of variables depends on the development stage of the product or process. Not as much is known about the product or process early in the development cycle, so such experiments tend to include more variables. This is an attempt to gain a better understanding of the overall experimental space. As more is known, experiments tend to include fewer variables but finer resolution is required.

The third section of this chapter is devoted to simple comparative experiments. These experiments have only one variable. The fourth section is devoted to multi-variable experiments in which matrix designs are utilized.

2.5. Determine the Sample Size and Sample Trials

The sample size depends on the values of α, β, and δ chosen by the experimenter. In general, the lower the α and β risks and the lower the δ difference to be resolved, the larger the sample size. The equation used to compute the sample size depends on the type of experiment being run. For illustrative purposes, consider the wire bonding experiment. Two populations are being compared and the variances of each population are known and equal. The null and alternative hypotheses are as follows.

H_0: $\mu_1 = \mu_2$

H_a: $\mu_1 \neq \mu_2$ (double-sided hypothesis)

The sample size (N) is found by the following equation.

$$N = 2\left(U_\alpha + U_\beta\right)^2 \frac{\sigma^2}{\delta^2} \tag{19.1}$$

where U_α is the normal distribution number for alpha risk and U_β is the normal distribution number for beta risk. Use Table 19.3 to find U_α when the alternative hypothesis is single-sided. Use Table 19.4 to find U_α when the alternative hypothesis is double-sided. U_β is obtained from Table 19.3 in all cases. For two-population experiments, N is the sample size needed from each population. As α and β risks decrease, the U_α and U_β values increase. Equation (19.1) also shows that as the variance increases, the required sample size increases.

Use Tables 19.3 and 19.4 when variances are known. These tables are based on the normal distribution. Use Tables 19.5 and 19.6 when the variances are unknown. In these cases, the variance must be estimated from previous work or estimated from the experimental results. Tables 19.5 and 19.6 are based on the t distribution.

After the sample size has been determined, the next step is to determine the sample trials. In the case of a single variable experiment such as the wire bonding example, this is straightforward. There are two populations: one at temperature, T_1, and another at temperature, T_2. Process N samples at T_1 and another N samples at T_2.

Multi-variable experiments are somewhat more complex. Matrix designs simplify determining the sample trials. Once the experimenter chooses the resolution of the experiment, the number of variables, and calculates the sample size, the proper matrix can be determined.

2.6. Calculate the Test Criterion

The test criterion is the value with which the sample means are compared to decide between the null and alternative hypotheses. The

Table 19.3. Probability Points of the Normal Distribution (Single-sided, σ^2 Known)

α or β	U
0.001	3.090
0.005	2.576
0.010	2.326
0.015	2.170
0.020	2.054
0.025	1.960
0.050	1.645
0.100	1.282
0.150	1.036
0.200	0.842
0.300	0.524
0.400	0.253
0.500	0.000

Table 19.4. Probability Points of the Normal Distribution (Double-sided, σ^2 Known)

α only	U
0.001	3.291
0.005	2.807
0.010	2.576
0.015	2.432
0.020	2.326
0.025	2.241
0.050	1.960
0.100	1.645
0.150	1.440
0.200	1.282
0.300	1.036
0.400	0.842
0.500	0.675

Table 19.5. Probability Points of the *t* Distribution
(Single-sided, σ^2 Unknown)

φ	α or β						
	0.005	**0.01**	**0.025**	**0.05**	**0.10**	**0.20**	**0.30**
1	63.66	31.82	12.71	6.31	3.08	1.38	0.73
2	9.93	6.97	4.30	2.92	1.89	1.06	0.62
3	5.84	4.54	3.18	2.35	1.64	0.98	0.58
4	4.60	3.75	2.78	2.13	1.53	0.94	0.57
5	4.03	3.37	2.57	2.02	1.48	0.92	0.56
6	3.71	3.14	2.45	1.94	1.44	0.91	0.56
7	3.50	3.00	2.37	1.90	1.42	0.90	0.55
8	3.36	2.90	2.31	1.86	1.40	0.90	0.55
9	3.25	2.82	2.26	1.83	1.38	0.89	0.54
10	3.17	2.76	2.23	1.81	1.37	0.89	0.54
15	2.95	2.60	2.13	1.75	1.34	0.87	0.54
20	2.85	2.53	2.09	1.73	1.33	0.86	0.53
25	2.79	2.49	2.06	1.71	1.32	0.86	0.53
30	2.75	2.46	2.04	1.70	1.31	0.85	0.53
60	2.66	2.39	2.00	1.67	1.30	0.85	0.53
120	2.62	2.36	1.98	1.66	1.29	0.85	0.53
∞	2.58	2.33	1.96	1.65	1.28	0.84	0.52

Table 19.6. Probability Points of the *t* Distribution
(Double-sided, σ^2 Unknown)

φ	α only						
	0.005	**0.01**	**0.02**	**0.05**	**0.10**	**0.20**	**0.30**
1	127.00	63.66	31.82	12.71	6.31	3.08	1.96
2	14.10	9.93	6.97	4.30	2.92	1.89	1.39
3	7.45	5.84	4.54	3.18	2.35	1.64	1.25
4	5.60	4.60	3.75	2.78	2.13	1.53	1.19
5	4.77	4.03	3.37	2.57	2.02	1.48	1.16
10	3.58	3.17	2.76	2.23	1.81	1.37	1.09
15	3.29	2.95	2.60	2.13	1.75	1.34	1.07
20	3.15	2.85	2.53	2.09	1.73	1.33	1.06
25	3.08	2.79	2.49	2.06	1.71	1.32	1.06
30	3.03	2.75	2.46	2.04	1.70	1.31	1.05
60	2.91	2.66	2.39	2.00	1.67	1.30	1.05
120	2.86	2.62	2.36	1.98	1.66	1.29	1.05
∞	2.81	2.58	2.33	1.96	1.65	1.28	1.04

equation used to calculate the test criterion depends on the type of experiment. The test criterion is designated by $\overline{X}^*$ when one population is involved and by $\left|\overline{X}_1 - \overline{X}_2\right|^*$ when two populations are compared. For the wire bond example where the variance is known and two populations are being compared, the test criterion is calculated by:

$$\left|\overline{X}_1 - \overline{X}_2\right|^* = U_\alpha \sigma \sqrt{\frac{1}{N_1} + \frac{1}{N_2}} \tag{19.2}$$

where $\overline{X}_1$ and $\overline{X}_2$ are the sample means from population 1 and 2, and N_1 and N_2 are the number of samples from population 1 and 2.

For experiments where the variance is unknown or cannot be estimated, calculate the test criterion after performing the experiment. In these cases, estimate the variance based on the experimental results in order to calculate the test criterion.

2.7. Final Steps

After the experiment has been planned correctly, the rest of the experiment steps are straightforward.

Perform the experiment

Perform the experiment by obtaining the correct number of samples for each trial condition. Test the samples for the parameter of interest, and compute the statistics for each sample group. If the variance is unknown, calculate the sample mean and variance. If the variance is already known, only the sample mean is necessary.

Compare sample means with test criterion

For experiments where one population is sampled, compare $\overline{X}$ with $\overline{X}^*$. For experiments where two populations are sampled, compare $\left|\overline{X}_1 - \overline{X}_2\right|$ with $\left|\overline{X}_1 - \overline{X}_2\right|^*$.

Make decision

The decision making step is automatic.

- If the sample mean is greater than or equal to the test criterion, accept the alternative hypothesis. There is at least $(1 - \alpha)(100)$ percent confidence this decision is correct.

- If the sample mean is less than the test criterion, accept the null hypothesis. This decision is correct with at least $(1 - \beta)(100)$ percent confidence.

3. SINGLE VARIABLE EXPERIMENTS— COMPARING MEANS

Now that we know the basics of how to design and perform an experiment, let's get down to the practical applications. This section covers single variable experiments—experiments where one variable is changed from one level to another, holding all other variables constant.

This section contains many cases of single variable experiments. There are different types of problems depending on whether the variance is known, unknown, or can be estimated from previous work. And there are different types depending on whether there is one population, two populations with equal variance, or two populations with unequal variance. Specifically, here are the cases that will be covered:

- Case 1a. $H_0: \mu_1 = \mu_0$; variance (σ_1^2) is known.

- Case 1b. $H_0: \mu_1 = \mu_2$; $\sigma_1^2 = \sigma_2^2$ and both are known.

- Case 1c. $H_0: \mu_1 = \mu_2$; $\sigma_1^2 \neq \sigma_2^2$ and both are known.

- Case 2a. $H_0: \mu_1 = \mu_0$; σ_1^2 is unknown.

- Case 2b. $H_0: \mu_1 = \mu_2$; $\sigma_1^2 = \sigma_2^2$ but both are unknown.

- Case 2c. $H_0: \mu_1 = \mu_2$; $\sigma_1^2 \neq \sigma_2^2$ but both are unknown.

- Cases 3a, b, c. Same as Cases 2a, b, c respectively except that an estimate of variance (S^2) is known from previous work.

The formulas for calculating sample sizes and test criterion are different depending on the type of experiment. Table 19.7 summarizes the sample size formulas, and Table 19.8 summarizes the test criterion formulas.

In addition, two special cases will be presented.

Table 19.7. Summary of Sample Size Formulas[51]

| | $H_0: \mu_1 = \mu_0$ | $H_0: \mu_1 = \mu_2$ | |
		$\sigma_1^2 = \sigma_2^2$	$\sigma_1^2 \neq \sigma_2^2$
σ^2 known	<u>Case 1a:</u> $N = (U_\alpha + U_\beta)^2 \dfrac{\sigma^2}{\delta^2}$	<u>Case 1b:</u> $N_1 = N_2 = 2(U_\alpha + U_\beta)^2 \dfrac{\sigma^2}{\delta^2}$	<u>Case 1c:</u> $N_1 = (U_\alpha + U_\beta)^2 \dfrac{\sigma_1(\sigma_1 + \sigma_2)}{\delta^2}$ $N_2 = (U_\alpha + U_\beta)^2 \dfrac{\sigma_2(\sigma_1 + \sigma_2)}{\delta^2}$
σ^2 and S^2 unknown	<u>Case 2a:</u> $N = (U_\alpha + U_\beta)^2 \dfrac{\sigma^2}{\delta^2}$ $\phi = N - 1$ $N_t = (t_\alpha + t_\beta)^2 \dfrac{\sigma^2}{\delta^2}$ State δ in terms of σ	<u>Case 2b:</u> $N_1 = N_2 = 2(U_\alpha + U_\beta)^2 \dfrac{\sigma^2}{\delta^2}$ $\phi = N_1 + N_2 - 2$ $N_{t1} = N_{t2} = 2(t_\alpha + t_\beta)^2 \dfrac{\sigma^2}{\delta^2}$ State δ in terms of σ	<u>Case 2c:</u> $N_1 = (U_\alpha + U_\beta)^2 \dfrac{\hat{\sigma}_1(\hat{\sigma}_1 + \hat{\sigma}_2)}{\delta^2}$ $N_2 = (U_\alpha + U_\beta)^2 \dfrac{\hat{\sigma}_2(\hat{\sigma}_1 + \hat{\sigma}_2)}{\delta^2}$ $\phi_1 = N_1 - 1, \ \phi_2 = N_2 - 1$ $N_{t1} = (t_\alpha + t_\beta)^2 \dfrac{\hat{\sigma}_1(\hat{\sigma}_1 + \hat{\sigma}_2)}{\delta^2}$ $N_{t2} = (t_\alpha + t_\beta)^2 \dfrac{\hat{\sigma}_2(\hat{\sigma}_1 + \hat{\sigma}_2)}{\delta^2}$
S^2 known	<u>Case 3a:</u> $N_t = (t_\alpha + t_\beta)^2 \dfrac{S^2}{\delta^2}$	<u>Case 3b:</u> $N_{t1} = N_{t2} = 2(t_\alpha + t_\beta)^2 \dfrac{S^2}{\delta^2}$	<u>Case 3c:</u> $N_{t1} = (t_\alpha + t_\beta)^2 \dfrac{S_1(S_1 + S_2)}{\delta^2}$ $N_{t2} = (t_\alpha + t_\beta)^2 \dfrac{S_2(S_1 + S_2)}{\delta^2}$

Table 19.8. Summary of Test Criterion Formulas[52]

	$H_0: \mu_1 = \mu_0$	$H_0: \mu_1 = \mu_2$	
		$\sigma_1^2 = \sigma_2^2$	$\sigma_1^2 \neq \sigma_2^2$
σ^2 **known**	<u>Case 1a:</u> $\bar{X}^* = \mu_0 + \dfrac{\sigma U_\alpha}{\sqrt{N}}$	<u>Case 1b:</u> $\left|\bar{X}_1 - \bar{X}_2\right|^* = U_\alpha \sigma \sqrt{\dfrac{1}{N_1} + \dfrac{1}{N_2}}$	<u>Case 1c:</u> $\left|\bar{X}_1 - \bar{X}_2\right|^* = U_\alpha \sqrt{\dfrac{\sigma_1^2}{N_1} + \dfrac{\sigma_2^2}{N_2}}$
σ^2 **and** S^2 **unknown**	<u>Case 2a:</u> $\bar{X}^* = \mu_0 + \dfrac{t_\alpha S}{\sqrt{N_t}}$ $S^2 = \dfrac{\sum\left(X_i - \bar{X}\right)^2}{N_t - 1}$	<u>Case 2b:</u> $\left|\bar{X}_1 - \bar{X}_2\right|^* = t_\alpha S \sqrt{\dfrac{1}{N_{t1}} + \dfrac{1}{N_{t2}}}$ $S^2 = \dfrac{\sum\left(X_{i1} - \bar{X}_1\right)^2 + \sum\left(X_{j2} - \bar{X}_2\right)^2}{N_{t1} + N_{t2} - 2}$	<u>Case 2c:</u> $\left|\bar{X}_1 - \bar{X}_2\right|^* = t_\alpha \sqrt{\dfrac{S_1^2}{N_{t1}} + \dfrac{S_2^2}{N_{t2}}}$ $S_1^2 = \dfrac{\sum\left(X_{i1} - \bar{X}_1\right)^2}{N_{t1} - 1}$ $S_2^2 = \dfrac{\sum\left(X_{j2} - \bar{X}_2\right)^2}{N_{t2} - 1}$
S^2 **known**	<u>Case 3a:</u> Same as Case 2a	<u>Case 3b:</u> Same as Case 2b	<u>Case 3c:</u> Same as Case 2c

- <u>Special Case 1.</u> Comparison of Paired Data
- <u>Special Case 2.</u> What To Do When Some Data Lost

3.1. Variance Is Known

For the cases where the variances are known, probabilities are based on the normal distribution.

3.1.1. Case 1a. Single Population, Variance Known

$$H_0: \mu_1 = \mu_0; \text{ variance } (\sigma_1^2) \text{ is known}$$

The wire bonding process of an integrated circuit (IC) involves bonding wires from the IC to the package. The population mean of the wire bond strength (μ_0) is 10.0 grams-force and the variance (σ_0^2) is known to be 1.5. This IC is experiencing failures during reliability testing for weak bond strength. It is thought that increasing the wire bonding temperature from 250°C to 300°C will increase the bond strength. An increase in bond strength of 2.0 grams-force is considered significant. The experimenter is willing to accept an α risk of 5% and a β risk of 10%.

- **Objective**

  $H_0: \mu_{300} = \mu_{250} = 10.0$ grams-force

  $H_a: \mu_{300} > 10.0$ grams-force (single-sided)

- **Choose α, β, and δ**

  $\alpha = 0.05$

  $\beta = 0.10$

  $\delta = 2$ grams-force $\delta^2 = 4$

  $\sigma^2 = 1.5$ $\sigma = 1.22$ grams-force

- **Calculate sample size**

  Look up values of U_α and U_β from Table 19.3 (single-sided normal distribution).

---

51. William J. Diamond, *Practical Experiment Designs for Engineers and Scientists, 2nd Edition,* New York: Van Nostrand Reinhold, 1989.

52. Ibid.

$$U_\alpha = 1.645$$

$$U_\beta = 1.282$$

Get the sample size formula from Table 19.7.

$$N = \left(U_\alpha + U_\beta\right)^2 \frac{\sigma^2}{\delta^2} = (1.645 + 1.282)^2 \frac{1.5}{4} = 3.21$$

Round N up to 4.

- **Calculate criterion**

  Get the formula from Table 19.8.

$$\overline{X}^* = \mu_0 + \frac{\sigma U_\alpha}{\sqrt{N}} = 10 + \frac{(1.22)(1.645)}{\sqrt{4}} = 11.01$$

- **Perform the experiment**

  Make 4 wire bonds with the temperature of the wire bonder set at 300°C, and test for the bond strength.

  Results: 11.3, 12.7, 9.8, and 10.9 grams-force

  $\overline{X}_{300} = 11.18$ grams-force

- **Compare $\overline{X}$ with $\overline{X}^*$**

$$\left(\overline{X} = 11.18\right) > \left(\overline{X}^* = 11.01\right)$$

- **Make decision**

  Accept H_a. With at least 95 percent confidence, it can be stated that the correct decision is to increase the bonding temperature from 250°C to 300°C.

3.1.2. Case 1b. Two Populations, Variances Known and Equal

$H_0: \mu_1 = \mu_2; \ \sigma_1^2 = \sigma_2^2$ and both are known

Enriching the oxygen concentration of air is of interest for enhanced combustion applications. This oxygen enrichment process can be accomplished by "filtering" air through polymer membranes.[53] Vendors A and B are being evaluated as suppliers of the polymer membranes. Both vendors state they meet the engineering specification which is to be able to produce a 40% oxygen (O_2) concentration for a given flow rate and pressure ratio across the membrane. They both state the variance (σ^2) is 4×10^{-4}. Each vendor quotes the same price, but vendor A is the current vendor. Thus there would be some costs associated with switching to vendor B. It is desirable to select the vendor with the highest oxygen concentration (μ). A difference of 4% is considered significant.

- **Objective**

    $H_0: \mu_A = \mu_B$

    $H_a: \mu_A \neq \mu_B$ (double-sided)

    The alternative hypothesis can also be stated:

    $H_{a1}: \mu_A > \mu_B$ with $\alpha/2$ risk

    $H_{a2}: \mu_B > \mu_A$ with $\alpha/2$ risk

- **Choose α, β, and δ and state σ^2**

    $\alpha = 0.05$

    $\beta = 0.10$

    $\delta = 4\%\ O_2 = 0.04$ $\delta^2 = 0.0016$

    $\sigma_A^2 = 0.0004$ $\sigma_A = 0.02 = 2\%\ O_2$

    $\sigma_B^2 = 0.0004$ $\sigma_B = 0.02 = 2\%\ O_2$

- **Calculate sample size**

    Look up the value of U_α from Table 19.4 (double-sided normal distribution). Look up the value of U_β from Table 19.3.

    $U_\alpha = 1.96$

    $U_\beta = 1.282$

    Compute N_A and N_B using the formula from Table 19.7.

---

53. Marcel Mulder, *Basic Principles of Membrane Technology,* Dordrecht, The Netherlands: Kluwer Academic Publishers, 1991.

$$N_A = N_B = 2\left(U_\alpha + U_\beta\right)^2 \frac{\sigma^2}{\delta^2} = 2(1.96 + 1.282)^2 \frac{(0.0004)}{(0.0016)} = 5.26$$

Round N_A and N_B up to 6.

- **Calculate criterion**

    Get the formula from Table 19.8.

$$\left|\overline{X}_A - \overline{X}_B\right|^* = U_\alpha \sigma \sqrt{\frac{1}{N_A} + \frac{1}{N_B}} = (1.96)(0.02)\sqrt{\frac{1}{6} + \frac{1}{6}} = 0.0226$$

- **Perform the experiment**

    Have each vendor supply 6 membrane samples, test the samples, and calculate $\overline{X}_A$ and $\overline{X}_B$.

    $\overline{X}_A = 49.6\% \; O_2 = 0.496$

    $\overline{X}_B = 46.6\% \; O_2 = 0.466$

    $\left|\overline{X}_A - \overline{X}_B\right| = 0.030$

- **Compare** $\left|\overline{X}_A - \overline{X}_B\right|$ **with** $\left|\overline{X}_A - \overline{X}_B\right|^*$

$$\left(\left|\overline{X}_A - \overline{X}_B\right| = 0.030\right) > \left(\left|\overline{X}_A - \overline{X}_B\right|^* = 0.0226\right)$$

- **Make decision**

    Since $\overline{X}_A > \overline{X}_B$ and $\left|\overline{X}_A - \overline{X}_B\right| > \left|\overline{X}_A - \overline{X}_B\right|^*$ accept H_{a1} with $[1 - (\alpha/2)](100)$ percent confidence. Accept H_{a1}: $\mu_A > \mu_B$ with 97.5% confidence. The conclusion is the O_2 concentration is significantly greater for the membranes from vendor A than from vendor B. Therefore, continue to purchase the membranes from vendor A.

3.1.3. Case 1c. Two Populations, Variances Known and Unequal

$H_0: \mu_1 = \mu_2; \sigma_1^2 \neq \sigma_2^2$ and both are known

For Case 1c, the example used is the same as Case 1b except that for Case 1c, vendor A says that the variance (σ^2) is 9×10^{-4}.

- **Objective**

  $H_0: \mu_A = \mu_B$

  $H_a: \mu_A \neq \mu_B$ (double-sided)

- **Choose α, β, and δ and state σ^2**

  $\alpha = 0.05$

  $\beta = 0.10$

  $\delta = 4\% \, O_2 = 0.04$ $\qquad$ $\delta^2 = 0.0016$

  $\sigma_A^2 = 0.0009$ $\qquad$ $\sigma_A = 0.03 = 3\% \, O_2$

  $\sigma_B^2 = 0.0004$ $\qquad$ $\sigma_B = 0.02 = 2\% \, O_2$

- **Calculate sample size**

  Look up the value of U_α from Table 19.4 (double-sided normal distribution). Look up the value of U_β from Table 19.3.

  $U_\alpha = 1.96$

  $U_\beta = 1.282$

  Compute N_A and N_B using the formulas from Table 19.7.

  $$N_A = \left(U_\alpha + U_\beta\right)^2 \frac{\sigma_A(\sigma_A + \sigma_B)}{\delta^2} = (1.96 + 1.282)^2 \frac{0.03(0.03 + 0.02)}{(0.0016)} = 9.85$$

  Round N_A up to 10.

  $$N_B = \left(U_\alpha + U_\beta\right)^2 \frac{\sigma_B(\sigma_A + \sigma_B)}{\delta^2} = (1.96 + 1.282)^2 \frac{0.02(0.03 + 0.02)}{(0.0016)} = 6.57$$

  Round N_B up to 7.

- **Calculate criterion**

  Get the formula from Table 19.8.

$$\left|\overline{X}_A - \overline{X}_B\right|^* = U_\alpha \sqrt{\frac{\sigma_A^2}{N_A} + \frac{\sigma_B^2}{N_B}} = 1.96 \sqrt{\frac{0.0009}{10} + \frac{0.0004}{7}} = 0.0238$$

- **Perform the experiment**

  Have vendor A supply 10 membrane samples and vendor B supply 7 membrane samples. Test the samples and calculate $\overline{X}_A$ and $\overline{X}_B$.

  $\overline{X}_A = 49.6\% \, O_2 = 0.496$

  $\overline{X}_B = 46.6\% \, O_2 = 0.466$

  $\left|\overline{X}_A - \overline{X}_B\right| = 0.030$

- **Compare** $\left|\overline{X}_A - \overline{X}_B\right|$ **with** $\left|\overline{X}_A - \overline{X}_B\right|^*$

  $$\left(\left|\overline{X}_A - \overline{X}_B\right| = 0.030\right) > \left(\left|\overline{X}_A - \overline{X}_B\right|^* = 0.0238\right)$$

- **Make decision**

  Accept H_{a1}: $\mu_A > \mu_B$ with 97.5% confidence. The conclusion is to purchase the membranes from vendor A.

3.2. Variance and Estimate of Variance Are Unknown

For the cases where the variance and the estimate of variance are unknown, probabilities are based on the t distribution. The t distribution introduces an additional degree of uncertainty and was portrayed earlier in Figure 18.2. The additional degree of uncertainty increases the number of samples required for a given α and β risk.

There are two other experimental design differences when the variance is unknown. One is the test criterion is calculated after the experiment rather than before. Calculate the number of samples, perform the experiment, and calculate the estimate of variance (S^2) based on the experiment results. Use this estimate of variance to calculate the test criterion. The other difference is the significant difference (δ) should be stated in units of σ rather than as an absolute number to simplify the sample size calculation.

3.2.1. Case 2a. Single Population, Variance Unknown

$$H_0: \mu_1 = \mu_0; \quad \text{variance } (\sigma_1^2) \text{ is unknown}$$

Case 2a is similar to Case 1a except the variance is not known. Thus for Case 2a, the δ will be expressed in terms of σ and the t distribution will be used instead of the normal distribution.

For this example, recall the wire bond pull strength of a process is 10.0 grams-force with the bonding temperature set at 250°C. This experiment is designed to see if the bond strength can be improved by one standard deviation by increasing the bonding temperature to 300°C.

- **Objective**

  $H_0: \mu_{300} = \mu_{250} = 10.0$ grams-force

  $H_a: \mu_{300} > 10.0$ grams-force (single-sided)

- **Choose α, β, and δ**

  $\alpha = 0.05$

  $\beta = 0.10$

  $\delta = \sigma \qquad\qquad \delta^2 = \sigma^2$

- **Calculate sample size**

  Look up values of U_α and U_β from Table 19.3 (single-sided normal distribution).

  $U_\alpha = 1.645$

  $U_\beta = 1.282$

  Get the sample size formula based on the normal distribution from Table 19.7.

  $$N = \left(U_\alpha + U_\beta\right)^2 \frac{\sigma^2}{\delta^2} = (1.645 + 1.282)^2 \frac{\sigma^2}{\sigma^2} = 8.57$$

  $\phi = N - 1 = 7.6$

  Look up t_α from Table 19.5.

  $t_\alpha = 1.88$ for $\alpha = 0.05$, $\phi = 7.6$ (interpolate)

  Look up t_β from Table 19.5.

  $t_\beta = 1.41$ for $\beta = 0.10$, $\phi = 7.6$ (interpolate)

Recalculate the sample size based on the t distribution. Get the formula from Table 19.7.

$$N_t = \left(t_\alpha + t_\beta\right)^2 \frac{\sigma^2}{\delta^2} = (1.88 + 1.41)^2 \frac{\sigma^2}{\sigma^2} = 10.82$$

Round N_t up to 11.

- **Perform the experiment**

  Make 11 wire bonds with the temperature of the wire bonder set at 300°C, and calculate $\overline{X}$ and S^2 from the sample data.

  $\overline{X}_{300} = 11.2$ grams-force

  Get the sample variance formula from Table 19.8.

$$S^2 = \frac{\sum\left(X_i - \overline{X}\right)^2}{N_t - 1} = 4.0 \qquad S = 2.0$$

- **Calculate criterion**

  Get the formula from Table 19.8.

  $t_\alpha = 1.81$ for $\alpha = 0.05$, $\phi = 10$ (from Table 19.5)

$$\overline{X}^* = \mu_0 + \frac{t_\alpha S}{\sqrt{N_t}} = 10 + \frac{(1.81)(2.0)}{\sqrt{11}} = 11.09$$

- **Compare $\overline{X}$ with $\overline{X}^*$**

$$\left(\overline{X} = 11.2\right) > \left(\overline{X}^* = 11.1\right)$$

- **Make decision**

  Accept H_a: $\mu_{300} > 10.0$ with at least 95 percent confidence. Increasing the bonding temperature to 300°C will produce wire bonds with a mean bond strength of greater than 10 grams-force.

3.2.2. Case 2b. Two Populations, Variances Unknown But Equal

H_0: $\mu_1 = \mu_2$; variances are equal and unknown

Case 2b is similar to Case 1b except that the variance is unknown. Suppose you are evaluating polymer membranes to see which one produces the higher O_2 concentration from air. A vendor proposes two types of membranes that are made from different formulations. The vendor is certain the variance will be the same for both membranes, but the variance value is unknown.

- **Objective**

  H_0: $\mu_{\text{Membrane A}} = \mu_{\text{Membrane B}}$

  H_a: $\mu_{\text{Membrane A}} \neq \mu_{\text{Membrane B}}$ (double-sided)

- **Choose α, β, and δ**

  $\alpha = 0.05$

  $\beta = 0.10$

  $\delta = \sigma$ $\qquad\qquad\qquad \delta^2 = \sigma^2$

- **Calculate sample size**

  Look up the value of U_α from Table 19.4 (double-sided normal distribution). Look up the value of U_β from Table 19.3.

  $U_\alpha = 1.96$

  $U_\beta = 1.282$

  Compute N_A and N_B. Get the sample size formula based on the normal distribution from Table 19.7.

  $$N_A = N_B = 2\left(U_\alpha + U_\beta\right)^2 \frac{\sigma^2}{\delta^2} = 2(1.96 + 1.282)^2 \frac{\sigma^2}{\sigma^2} = 21.0$$

  Compute ϕ.

  $\phi = N_A + N_B - 2 = 21 + 21 - 2 = 40$

  Look up t_α by interpolating from Table 19.6.

  $t_\alpha = 2.03$ for $\alpha = 0.05$, $\phi = 40$

  Look up t_β by interpolating from Table 19.5.

$t_\beta = 1.31$ for $\beta = 0.10$, $\phi = 40$

Recalculate the sample size based on the t distribution. Get the formula from Table 19.7.

$$N_{tA} = N_{tB} = 2(t_\alpha + t_\beta)^2 \frac{\sigma^2}{\delta^2} = 2(2.03 + 1.31)^2 \frac{\sigma^2}{\sigma^2} = 22.3$$

- **Perform the experiment**

  Have the vendor supply 22 membrane samples of each formulation, and test the samples.

  $\overline{X}_A = 42.3\% \ O_2 = 0.423$

  $\overline{X}_B = 43.9\% \ O_2 = 0.439$

  $\left| \overline{X}_A - \overline{X}_B \right| = 0.016$

  Get the sample variance formula from Table 19.8.

  $$S^2 = \frac{\sum \left(X_{iA} - \overline{X}_A \right)^2 + \sum \left(X_{jB} - \overline{X}_B \right)^2}{N_{tA} + N_{tB} - 2} = 9 \times 10^{-4}$$

  $S = 0.03$, $\phi = 42$

- **Calculate criterion**

  Get the formula from Table 19.8.

  $t_\alpha = 2.02$ for $\alpha = 0.05$, $\phi = 42$ (from Table 19.6)

  $$\left| \overline{X}_A - \overline{X}_B \right|^* = t_\alpha S \sqrt{\frac{1}{N_{tA}} + \frac{1}{N_{tB}}} = (2.02)(0.03)\sqrt{\frac{1}{22} + \frac{1}{22}} = 0.0183$$

- **Compare** $\left| \overline{X}_A - \overline{X}_B \right|$ **with** $\left| \overline{X}_A - \overline{X}_B \right|^*$

  $$\left(\left| \overline{X}_A - \overline{X}_B \right| = 0.016 \right) < \left(\left| \overline{X}_A - \overline{X}_B \right|^* = 0.0183 \right)$$

- **Make decision**

Accept H_0: $\mu_{\text{Membrane A}} = \mu_{\text{Membrane B}}$ with at least 90% confidence. The difference in the resulting O_2 concentrations between membrane A and membrane B was found to be insignificant.

3.2.3. Case 2c. Two Populations, Variances Unknown But Unequal

H_0: $\mu_1 = \mu_2$; variances are not equal and unknown

Case 2c is somewhat similar to Case 1c except that the variances are not known. Suppose again you are evaluating polymer membranes to see which membrane produces the highest O_2 concentration from air. A vendor proposes two types of membranes that are made of different formulations. The vendor estimates membrane A will have a variance of about 2.5×10^{-3} and membrane B will have a variance of about 9×10^{-4}. These estimates of variance are not based on data. The specification for the part calls for a population mean O_2 concentration of 40% or greater.

- **Objective**

  H_0: $\mu_{\text{Membrane A}} = \mu_{\text{Membrane B}}$

  H_a: $\mu_{\text{Membrane A}} \neq \mu_{\text{Membrane B}}$ (double-sided)

- **Choose α, β, and δ and state $\hat\sigma_A^2$ and $\hat\sigma_B^2$ where $\hat\sigma^2$ means an estimate of σ^2 not based on data.**

  $\alpha = 0.05$

  $\beta = 0.10$

  $\delta = 4\%\ O_2 = 0.04$ $\qquad \delta^2 = 0.0016$

  $\hat\sigma_A^2 = 0.0025$ $\qquad \hat\sigma_A = 0.05$

  $\hat\sigma_B^2 = 0.0009$ $\qquad \hat\sigma_B = 0.03$

- **Calculate sample size**

  Look up the value of U_α from Table 19.4 (double-sided normal distribution). Look up the value of U_β from Table 19.3.

  $U_\alpha = 1.96$

  $U_\beta = 1.282$

  Compute N_A and N_B. Get the sample size formula based on the normal distribution from Table 19.7.

$$N_A = (U_\alpha + U_\beta)^2 \frac{\hat{\sigma}_A(\hat{\sigma}_A + \hat{\sigma}_B)}{\delta^2} = (1.96 + 1.282)^2 \frac{0.05(0.05 + 0.03)}{(0.0016)} = 26.3$$

$$N_B = (U_\alpha + U_\beta)^2 \frac{\hat{\sigma}_B(\hat{\sigma}_A + \hat{\sigma}_B)}{\delta^2} = (1.96 + 1.282)^2 \frac{0.03(0.05 + 0.03)}{(0.0016)} = 15.8$$

$\phi_A = 25.3$, $\phi_B = 14.8$

Recalculate the sample sizes based on the t distribution. Get the formula from Table 19.7. Get t_α from Table 19.6 and t_β from Table 19.5.

$$N_{tA} = (t_\alpha + t_\beta)^2 \frac{\hat{\sigma}_A(\hat{\sigma}_A + \hat{\sigma}_B)}{\delta^2} = (2.06 + 1.32)^2 \frac{0.05(0.05 + 0.03)}{0.0016} = 28.6$$

Round off to 29.

$$N_{tB} = (t_\alpha + t_\beta)^2 \frac{\hat{\sigma}_B(\hat{\sigma}_A + \hat{\sigma}_B)}{\delta^2} = (2.13 + 1.34)^2 \frac{0.03(0.05 + 0.03)}{0.0016} = 18.1$$

Round off to 18.

- **Perform the experiment**

   Have the vendor supply 29 samples of membrane A and 18 samples of membrane B. Test the samples.

   $\overline{X}_A = 45.9\% \, O_2 = 0.459$

   $\overline{X}_B = 49.6\% \, O_2 = 0.496$

   $\left| \overline{X}_A - \overline{X}_B \right| = 0.037$

   Get the sample variance formulas from Table 19.8.

$$S_A^2 = \frac{\sum \left(X_{iA} - \overline{X}_A \right)^2}{N_{tA} - 1} = 0.003$$

$$S_B^2 = \frac{\sum \left(X_{jB} - \overline{X}_B \right)^2}{N_{tB} - 1} = 0.0008$$

- **Calculate criterion**

  Get the formula from Table 19.8.

  $$\left|\overline{X}_A - \overline{X}_B\right|^* = t_\alpha \sqrt{\frac{S_A^2}{N_{tA}} + \frac{S_B^2}{N_{tB}}}$$

  $\phi = N_A + N_B - 2 = 29 + 18 - 2 = 45$

  $t_\alpha = 2.02$ (interpolating from Table 19.6)

  $$\left|\overline{X}_A - \overline{X}_B\right|^* = (2.02)\sqrt{\frac{0.003}{29} + \frac{0.0008}{18}} = 0.0246$$

- **Compare** $\left|\overline{X}_A - \overline{X}_B\right|$ **with** $\left|\overline{X}_A - \overline{X}_B\right|^*$

  $$\left(\left|\overline{X}_A - \overline{X}_B\right| = 0.037\right) > \left(\left|\overline{X}_A - \overline{X}_B\right|^* = 0.0246\right)$$

- **Make decision**

  Accept H_a: $\mu_{Membrane\ A} > \mu_{Membrane\ B}$ with at least 97.5% confidence. Conclusion: use membrane B.

3.3. Estimate of Variance Is Known from Previous Work

In Cases 2a, b, and c, the variance was not known. In Cases 2a and 2b, the significant difference (δ) needed to be stated in terms of the standard deviation (σ). Experiments were performed, and a value for S^2 was obtained from the experiment. This S^2 value was then used to calculate the criterion for testing the null hypothesis.

In some instances, the value of S^2 is known from previous work. This information can be used to reduce the amount of work and to improve the confidence level of the decision. When S^2 is known from previous work, the criterion can be computed before the experiment is performed. Cases 3a, b, and c are the same as Cases 2a, b, and c, respectively, except that an estimate of variance (S^2) is known from previous work. As shown in the summary tables of formulas (Tables

19.7 and 19.8), only the sample size formulas are different. The probabilities are again based on the t distribution.

3.3.1. Case 3a. Single Population, Estimate of Variance Known

H_0: $\mu_1 = \mu_0$; value of S^2 known from previous work

Since the Case 3 family is so similar to the Case 2 family, only one example (Case 3a) will be shown. Case 3a will be a continuation of Case 2a. In Case 2a, it was desirable to improve the wire bonding strength, and the effect of increasing the bonding temperature from 250°C to 300°C was observed. Eleven samples were produced with the temperature at 300°C, and it was found that $\overline{X}_{300} = 11.2$ grams-force. Also it was found $S^2 = 4.0$ for $\phi = 10$. It was concluded that $\mu_{300} > 10.0$ with at least 95% confidence.

After this experiment was completed, it was proposed that the bond strength may be increased further by increasing the bonding temperature to 350°C. For this case, σ^2 is not known, but S^2 is known with $\phi = 10$. This value of S^2 can be used to calculate N and $\overline{X}^*$ for investigating the bonding temperature at 350°C.

- **Objective**

  H_0: $\mu_{350} = \mu_{300} = 11.2$ grams-force

  H_a: $\mu_{350} > 11.2$ grams-force (single-sided)

- **Choose α, β, and δ and state S^2**

  $\alpha = 0.05$

  $\beta = 0.10$

  $\delta = 2$ grams-force $\quad \delta^2 = 4$

  $S^2 = 4, \phi = 10$ $\quad\quad S = 2$

- **Calculate sample size**

  Look up value of t_α from Table 19.5 (single-sided).

  $t_\alpha = 1.81$

  Look up t_β from Table 19.5.

  $t_\beta = 1.37$

Get the sample size formula from Table 19.7.

$$N_t = \left(t_\alpha + t_\beta\right)^2 \frac{S^2}{\delta^2} = (1.81 + 1.37)^2 \frac{4}{4} = 10.1$$

Round off to 10.

- **Calculate criterion**

  Get the formula from Table 19.8.

  $$\overline{X}^* = \mu_0 + \frac{t_\alpha S}{\sqrt{N_t}} = 11.2 + \frac{(1.81)(2.0)}{\sqrt{10}} = 12.35$$

- **Perform the experiment**

  Make 10 wire bonds with the temperature of the wire bonder set at 350°C and test.

  $$\overline{X}_{350} = 11.9 \text{ grams-force}$$

- **Compare $\overline{X}$ with $\overline{X}^*$**

  $$\left(\overline{X}_{350} = 11.9\right) < \left(\overline{X}^* = 12.35\right)$$

- **Make decision**

  Accept H_0. With at least 90% confidence, it can be stated that increasing the temperature to 350°C does not increase the bond strength by 2 grams-force.

3.4. Special Case 1: Comparison of Paired Data

There are some experiments where the results are paired. In other words the results from both population 1 and population 2 can be obtained from the same samples. Taking advantage of paired data is one of the most powerful techniques for reducing the required sample size. Pairing can be performed when measuring the effect of an extra processing treatment on population 1. Measure the same samples before and after the extra treatment, and the experiment data are the differences between these measurements. Since the standard deviation of the difference (σ_{diff}) is unknown, state the significant difference (δ_{diff}) in terms of the standard deviation in order to calculate the sample size.

Perform the experiment, and calculate the differences. Use the variance of the differences to calculate a test criterion.

3.4.1. Example of Paired Data

This is a special case of Case 2a. With the use of CCDs (charge-coupled devices) in imaging systems, it is desirable to have CCDs with the highest possible quantum efficiency. The higher the quantum efficiency, the more effective the imaging system is at producing images under low light conditions. The average quantum efficiency of the CCDs is 50% at a specified wavelength for a particular manufacturer. It is proposed the quantum efficiency can be increased by coating the CCDs with an anti-reflective film.

- **Objective**

  H_0: μ_{diff} (= $\mu_{with\ film}$ - $\mu_{without\ film}$) = 0

  H_a: μ_{diff} > 0 (single-sided)

- **Choose α, β, and δ**

  $\alpha = 0.05$

  $\beta = 0.10$

  $\delta_{diff} = \sigma_{diff}$

- **Calculate sample size**

  Look up values of U_α and U_β from Table 19.3 (single-sided normal distribution).

  $U_\alpha = 1.645$

  $U_\beta = 1.282$

  Get the sample size formula based on the normal distribution from Table 19.7.

  $$N_{pairs} = \left(U_\alpha + U_\beta\right)^2 \frac{\sigma^2}{\delta^2} = (1.645 + 1.282)^2 \frac{\sigma^2}{\sigma^2} = 8.6$$

  $\phi = N - 1 = 7.6$

  Recalculate the sample size based on the t distribution. Get the formula from Table 19.7.

  $t_\alpha = 1.88$ (interpolating from Table 19.5)

$t_\beta = 1.41$ (interpolating from Table 19.5)

$$N_t = \left(t_\alpha + t_\beta\right)^2 \frac{\sigma^2}{\delta^2} = (1.88 + 1.41)^2 \frac{\sigma^2}{\sigma^2} = 10.8$$

Round N_t up to 11.

- **Perform the experiment**

    Select 11 CCDs at random. Test the quantum efficiency (QE) of each CCD before and after the coating and calculate the difference.

Pair	QE (before coating)	QE (after coating)	Difference
1	0.52	0.67	0.15
2	0.55	0.62	0.07
3	0.46	0.56	0.10
4	0.42	0.54	0.12
5	0.57	0.74	0.17
6	0.52	0.52	0.00
7	0.50	0.55	0.05
8	0.58	0.70	0.12
9	0.43	0.53	0.10
10	0.49	0.64	0.15
11	0.50	0.58	0.08
Mean			**0.101**
Estimate of standard deviation			**0.0495**

$\overline{X}_{\text{diff}} = 0.101$

$S_{\text{diff}} = 0.0495$

- **Calculate criterion**

    Get the formula from Table 19.8.

    $\phi = N_{\text{diff}} - 1 = 11 - 1 = 10$

    $t_\alpha = 1.81$ (from Table 19.5)

    $$\overline{X}_{\text{diff}}^* = \frac{t_\alpha S_{\text{diff}}}{\sqrt{N_{\text{diff}}}} = \frac{(1.81)(0.0495)}{\sqrt{11}} = 0.027$$

- **Compare $\overline{X}_{\text{diff}}$ with $\overline{X}_{\text{diff}}^*$**

$$\left(\overline{X}_{\text{diff}} = 0.101\right) > \left(\overline{X}^{*}_{\text{diff}} = 0.027\right)$$

- **Make decision**

  Accept H_a: $\mu_{\text{diff}} > 0$ with at least 95% confidence. Change the process to include the anti-reflective film.

3.5. Special Case 2: What To Do When Some Data Lost

Special Case 2 involves the situation where:

1) it is not possible to obtain the number of samples calculated in the experimental design, or

2) during the process of testing the samples, some data were lost or obviously bad.

Take the example of Case 2c (Section 3.2.3.): $N_{tA} = 29$ and $N_{tB} = 18$. Suppose the vendor was only able to supply 15 samples of membrane A and 17 samples of membrane B. Or the vendor was able to supply $N_{tA} = 29$ and $N_{tB} = 18$, but only 15 valid data points for membrane A and 17 valid data points for membrane B were obtained.

In these situations, proceed as if the actual sample size obtained is the proper sample size. However the criterion must be recalculated based on the actual sample size.

- **Recalculate criterion**

  Get the formula from Table 19.8.

  $$\phi = N_A + N_B - 2 = 15 + 17 - 2 = 30$$

  $$t_\alpha = 2.04 \quad \text{for } \alpha = 0.05 \text{ and } \phi = 30 \text{ (from Table 19.6)}$$

  $$\left|\overline{X}_A - \overline{X}_B\right|^* = t_\alpha \sqrt{\frac{S_A^2}{N_A} + \frac{S_B^2}{N_B}} = (2.04)\sqrt{\frac{0.003}{15} + \frac{0.0008}{17}} = 0.0321$$

  This criterion is larger than the 0.0246 calculated had the desired sample size been obtained.

- **Compare** $\left|\overline{X}_A - \overline{X}_B\right|$ **with** $\left|\overline{X}_A - \overline{X}_B\right|^*$

If the results are still $\overline{X}_A = 0.459$ and $\overline{X}_B = 0.496$, then $\left|\overline{X}_A - \overline{X}_B\right| = 0.037$.

$$\left(\left|\overline{X}_A - \overline{X}_B\right| = 0.037\right) > \left(\left|\overline{X}_A - \overline{X}_B\right|^* = 0.0321\right)$$

- **Make decision**

    The decision would still be to accept H_a: $\mu_B > \mu_A$ and the confidence level would still be 97.5% or greater.

    However if $\left|\overline{X}_A - \overline{X}_B\right| = 0.030$ for instance, the decision would be to accept H_0: $\mu_A = \mu_B$. The confidence in this decision would be less than the 90% that was desired when the experiment was designed with $\beta = 0.10$.

4. MULTI-VARIABLE EXPERIMENTS

The previous section described how to set up experiments when there is one variable. This section describes how to set up experiments when there are multiple variables. The focus is on 2-level experiments—each variable has a low and high level. The ultimate tools to use for 2-level, multi-variable experiments are matrix designs. However, it is necessary to cover some general principles before explaining these designs.

4.1. General Principles

4.1.1. Interactions

For single-variable experiments, the engineer is concerned with the effect this variable has on the response of interest and whether or not this effect is significant. However for multi-variable experiments, there is a new dimension to consider: interactions between variables.

First consider the case of non-interacting variables. Figure 19.1 shows a two-variable plot of non-interacting variables where the variables are temperature and voltage. Whether the voltage is set at 5 V or 6 V, an increase in temperature from 80° to 100° results in a decrease in response of 200 units. Whether the temperature is set at 80° or 100°, an increase in voltage from 5 V to 6 V results in an increase in

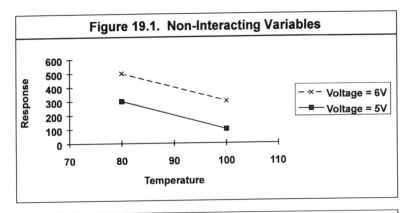

Figure 19.1. Non-Interacting Variables

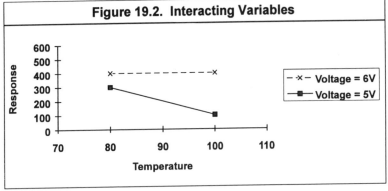

Figure 19.2. Interacting Variables

the response of 200 units. The effects of the two variables are independent. The response lines of non-interacting variables are parallel in a two-variable plot.

On the other hand, consider the case of interacting variables. Figure 19.2 shows an example where two variables are interacting. The effects of the variables are not independent. The effects must be stated in terms of both variables. For instance, when the voltage is 6 V, temperature has no effect on the response. However when the voltage is 5 V, increasing the temperature from 80° to 100° decreases the response by 200 units.

In fact, interactions can be detected with two-variable plots whenever the response lines are not parallel. The effect of an interaction must be greater than the test criterion in order to be significant. It is also possible there may be higher order interactions such as three-variable interactions and four-variable interactions. However in practice, three-variable interactions are much less likely than two-variable interactions, and four-variable interactions are much less likely than three-variable interactions. For most situations, be concerned mainly with two-variable interactions.

4.1.2. One-Variable-at-a-Time Experiments

A common approach to experimentation when more than one variable is involved is to perform simple comparative experiments for one variable at a time. For example, suppose there are three variables (A, B, and C) and the number of samples required ($N_{high} = N_{low}$) is calculated to be 4 for a given σ^2, α, β, and δ. N_{high} is the number of samples for the high level of each variable, and N_{low} is the number of samples for the low level.

Using the one-variable-at-a-time approach requires 24 samples.

First experiment set: determine effect of A keeping B and C constant

low A, low B, low C	4 samples
high A, low B, low C	4 samples

Second experiment set: determine effect of B keeping A and C constant

best A, low B, low C	4 samples
best A, high B, low C	4 samples

Third experiment set: determine effect of C keeping A and B constant

best A, best B, low C	4 samples
best A, best B, high C	4 samples

Total = 24 samples

There are two problems with one-variable-at-a-time experiments. The first problem is they do not consider the effect of any interactions. This is the biggest source of erroneous results for multi-variable experiments. The second problem is one-variable-at-a-time experiments are terribly inefficient. A full factorial design requires only 8 samples compared to 24 for the one-variable-at-a-time experiment.

The bottom line is never use the one-variable-at-a-time approach for multi-variable experiments.

4.1.3. Full Factorial Experiments

A full factorial experiment tests for all possible combinations of the variables at each level. For a two-level experiment with N variables, the number of combinations is 2^N. Thus for the previous example with three variables, there are $2^3 = 8$ combinations.

One way to show the trials of the matrix is to show the different combinations in a table by using "-" to designate the low level of a variable and "+" to designate the high level of a variable. A two-level, three-variable experiment can be arranged as follows.

Trial	A	B	C
1	-	-	-
2	+	-	-
3	-	+	-
4	+	+	-
5	-	-	+
6	+	-	+
7	-	+	+
8	+	+	+

Another way to show this example is to place the trial combinations at the corners of a cube as shown in Figure 19.3. Each variable represents a different dimension, and each dimension has a low and high value. Each corner of the cube represents a unique trial combination. The trial number for each combination is shown in parentheses.

Now let's see how the effects of the variables are determined. Suppose the experimenter makes one sample for each of the combinations. The difference between Trial 2 and Trial 1 is the level of A except for experimental error. The same can be said for Trials 4 and 3, Trials 6 and 5, and Trials 8 and 7. Each of these trial pairs can be used to determine the effect of variable A. Thus from these eight samples, the effect of A is measured four times, and this matches the number of samples required.

$$\left(\overline{X}_{high\ A} - \overline{X}_{low\ A} \right) = \frac{(T2 - T1) + (T4 - T3) + (T6 - T5) + (T8 - T7)}{4}$$

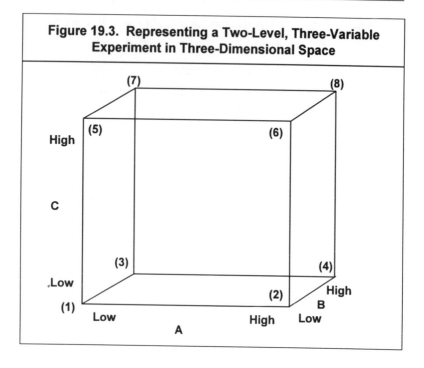

Figure 19.3. Representing a Two-Level, Three-Variable Experiment in Three-Dimensional Space

$$\left(\overline{X}_{\text{high A}} - \overline{X}_{\text{low A}}\right) = \frac{(T2 + T4 + T6 + T8)}{4} - \frac{(T1 + T3 + T5 + T7)}{4}$$

where Tn represents the nth trial.

The real benefit from using a matrix design is these same eight trials can be used to estimate the effect of the other variables. The effect of variable B can be estimated four times by examining the difference between trial pairs 3 and 1, 4 and 2, 7 and 5, and 8 and 6. Similarly the effect of variable C can be estimated by examining the difference between trial pairs 5 and 1, 6 and 2, 7 and 3, and 8 and 4.

$$\left(\overline{X}_{\text{high B}} - \overline{X}_{\text{low B}}\right) = \frac{(T3 + T4 + T7 + T8)}{4} - \frac{(T1 + T2 + T5 + T6)}{4}$$

$$\left(\overline{X}_{\text{high C}} - \overline{X}_{\text{low C}}\right) = \frac{(T5 + T6 + T7 + T8)}{4} - \frac{(T1 + T2 + T3 + T4)}{4}$$

4.1.3.1. Full Factorial Experiment Example

Case 1b. H_0: $\mu_{high} = \mu_{low}$; variances are known and equal

A chemical reaction consists of controlling three variables: the amount of catalyst at 1%, the temperature at 25°C, and the time at 30 minutes. This process is currently yielding the desired chemical with a population mean (μ_0) of 0.05 and a population variance (σ^2) of 0.0001. It is desirable to increase the yield. To increase yield, it is thought a catalyst amount of 2%, a temperature of 40°C, and a time of 60 minutes could increase the yield by 0.025 which would be a significant improvement.

- **Objective**

  Define the variables and the levels for each.

Variable	-	+
A: Catalyst amount	1%	2%
B: Temperature	25°C	40°C
C: Time	30 min.	60 min.

$(H_0)_A$: $\mu_{high\ A} = \mu_{low\ A}$, or $(\mu_{high\ A} - \mu_{low\ A}) = 0$

$(H_a)_A$: $\mu_{high\ A} > \mu_{low\ A}$, or $(\mu_{high\ A} - \mu_{low\ A}) > 0$ (single-sided)

$(H_0)_B$: $\mu_{high\ B} = \mu_{low\ B}$

$(H_a)_B$: $\mu_{high\ B} > \mu_{low\ B}$

$(H_0)_C$: $\mu_{high\ C} = \mu_{low\ C}$

$(H_a)_C$: $\mu_{high\ C} > \mu_{low\ C}$

- **Choose α, β, and δ and state σ^2**

  $\alpha = 0.05$

  $\beta = 0.05$

  $\delta = 2.5 \times 10^{-2} = 2.5\%$ $\delta^2 = 6.25 \times 10^{-4}$

  $\sigma^2 = 1 \times 10^{-4}$ $\sigma = 1 \times 10^{-2} = 1\%$

- **Calculate sample size**

  Get the formula from Table 19.7.

$$N_{high} = N_{low} = 2(U_\alpha + U_\beta)^2 \frac{\sigma^2}{\delta^2}$$

$U_\alpha = 1.645$ (from Table 19.3, single-sided)

$U_\beta = 1.645$ (from Table 19.3)

$$N_{high} = N_{low} = 2(1.645 + 1.645)^2 \frac{\left(1 \times 10^{-4}\right)}{\left(6.25 \times 10^{-4}\right)} = 3.5$$

Round off to 4.

A design of 4 samples at the high level and 4 samples at the low level for each variable is required. A $2^3 = 8$ trial experiment is adequate.

- **Define trials**

Trial	A	B	C
1	-	-	-
2	+	-	-
3	-	+	-
4	+	+	-
5	-	-	+
6	+	-	+
7	-	+	+
8	+	+	+

- **Calculate criterion**

  Get the formula from Table 19.8.

$$\left|\overline{X}_{high} - \overline{X}_{low}\right|^* = U_\alpha \sigma \sqrt{\frac{1}{N_{high}} + \frac{1}{N_{low}}} = (1.645)(1\%)\sqrt{\frac{1}{4} + \frac{1}{4}} = 1.16\%$$

- **Perform the experiment**

  Conduct a chemical reaction for each trial and measure the yield in %.

Trial	A	B	C	Yield (%)
1	1%	25°C	30 min.	5.0
2	2%	25°C	30 min.	5.2
3	1%	40°C	30 min.	8.0
4	2%	40°C	30 min.	7.8
5	1%	25°C	60 min.	4.5
6	2%	25°C	60 min.	4.8
7	1%	40°C	60 min.	8.4
8	2%	40°C	60 min.	7.5

Multiply each result by the treatment vectors to facilitate determining the effect of each variable.

Trial	Yield (%)	A	B	C
1	5.0	-5.0	-5.0	-5.0
2	5.2	+5.2	-5.2	-5.2
3	8.0	-8.0	+8.0	-8.0
4	7.8	+7.8	+7.8	-7.8
5	4.5	-4.5	-4.5	+4.5
6	4.8	+4.8	-4.8	+4.8
7	8.4	-8.4	+8.4	+8.4
8	7.5	+7.5	+7.5	+7.5
	Total	-0.6	+12.2	-0.8

$$\overline{X}_{high\ A} - \overline{X}_{low\ A} = \frac{-0.6\%}{4} = -0.15\%$$

$$\overline{X}_{high\ B} - \overline{X}_{low\ B} = \frac{12.2\%}{4} = 3.05\%$$

$$\overline{X}_{high\ C} - \overline{X}_{low\ C} = \frac{-0.8\%}{4} = -0.2\%$$

- **Compare results with the criterion**

$$\left(\overline{X}_{low\ A} - \overline{X}_{high\ A} = 0.15\%\right) < 1.16\%$$

Accept $(H_0)_A$: $\mu_{high\ A} = \mu_{low\ A}$

$$\left(\overline{X}_{high\ B} - \overline{X}_{low\ B} = 3.05\%\right) > 1.16\%$$

Accept $(H_a)_B$: $\mu_{high\ B} > \mu_{low\ B}$

$$\left(\overline{X}_{low\ C} - \overline{X}_{high\ C} = 0.2\%\right) < 1.16\%$$

Accept $(H_0)_C$: $\mu_{high\ C} = \mu_{low\ C}$

- **Make decision**

> Change the temperature of the chemical reaction to 40°C. This will increase yield by an estimated 3.05%. Increasing the catalyst amount or increasing the time will not increase yield.

In this last example, $N_{high} = N_{low} = 4$. Making one sample for each combination of a three-variable, full factorial matrix satisfied this sample size requirement. If the number of samples required is higher, the matrix can be replicated as many times as necessary in order to obtain the required sample size. For example if $N_{high} = N_{low} = 8$, obtain two samples for each treatment combination, or if $N_{high} = N_{low} = 16$, obtain four samples for each treatment combination.

The purpose of describing full factorial experiments was to introduce some basic concepts of matrix designs and show how they are superior to one-variable-at-a-time experiments. Determining the effects of interactions has not been covered yet. The rest of the chapter will be devoted to other matrix designs which can be even more efficient than full factorial designs. This matrix presentation will include how the effects of interactions are quantified. You should use these matrices when designing experiments to achieve maximum efficiency and effectiveness.

4.2. Matrix Designs

Hadamard matrices were discovered by a French mathematician, Jacques Hadamard. Using these matrices for experimental design was first described by Plackett and Burman.[54] Thus these designs are sometimes called Plackett-Burman designs. More applications were discussed by Hedayat and Wallis,[55] and later Diamond[56] showed their expanded usefulness for two-level, multi-variable designs.

---

54. R.L. Plackett, and J.P. Burman, "The Design of Optimum Multifactorial Experiments," *Biometrika*, 1946, **33**, 305-325.

55. A. Hedayat, and W.D. Wallis, "Hadamard Matrices and Their Applications," *The Annals of Statistics*, 1978, **Vol. 6, No. 6,** 1184-1238.

56. William J. Diamond, *Practical Experiment Designs for Engineers and Scientists, 2nd Edition,* New York: Van Nostrand Reinhold, 1989.

This section will describe three sizes of matrices: order 8, 16, and 32 Hadamard matrices. It is possible to include more variables, more trials, and detect greater resolution as the matrix size increases.

The procedure for designing a two-level multi-variable experiment is the same as outlined in Table 19.1.

- Define the objective. This includes deciding the number of variables and stating the null and alternate hypotheses for each variable.

- Choose the appropriate risks and significant difference. α, β, and δ

- Calculate the sample size. There are only two possible formulas.

$$N_{high} = N_{low} = 2\left(U_\alpha + U_\beta\right)^2 \frac{\sigma^2}{\delta^2} \qquad \text{if } \sigma^2 \text{ is known.}$$

$$N_{high} = N_{low} = 2\left(t_\alpha + t_\beta\right)^2 \frac{\sigma^2}{\delta^2} \qquad \text{if } \sigma^2 \text{ is unknown.}$$

- Calculate the test criterion. Again there are only two possible formulas.

$$\left|\overline{X}_{high} - \overline{X}_{low}\right|^* = U_\alpha \sigma \sqrt{\frac{1}{N_{high}} + \frac{1}{N_{low}}} \quad \text{if } \sigma^2 \text{ is known.}$$

$$\left|\overline{X}_{high} - \overline{X}_{low}\right|^* = t_\alpha S \sqrt{\frac{1}{N_{high}} + \frac{1}{N_{low}}} \quad \text{if } \sigma^2 \text{ is unknown.}$$

- Determine how the samples will be processed. This involves selecting which Hadamard matrix to use.

First select the desired resolution:

  - Resolution V: all main variable effects and all two-variable interactions can be determined. Assume all three-variable and higher order interactions are insignificant.

  - Resolution IV: all main variable effects and groups of two-variable interactions can be

Table 19.9. Summary of Matrix Designs			
Number of Variables	$N_{high} = N_{low}$	Resolution	Order of Matrix
2	≤ 4	V	8
2	≤ 8	V	16
2	≤ 16	V	32
3	≤ 4	V	8
3	≤ 8	V	16
3	≤ 16	V	32
4	≤ 4	IV	8
4	≤ 8	V	16
4	≤ 16	V	32
5	≤ 4	III	8
5	≤ 8	V	16
5	≤ 16	V	32
6	≤ 4	III	8
6	≤ 8	IV	16
6	≤ 16	V	32
7	≤ 4	III	8
7	≤ 8	IV	16
7	≤ 16	IV	32
8	≤ 8	IV	16
8	≤ 16	IV	32
9	≤ 8	III	16
9	≤ 16	IV	32
10	≤ 8	III	16
10	≤ 16	IV	32
11	≤ 8	III	16
11	≤ 16	IV	32
12	≤ 8	III	16
12	≤ 16	IV	32
13	≤ 8	III	16
13	≤ 16	IV	32
14	≤ 8	III	16
14	≤ 16	IV	32
15	≤ 8	III	16
15	≤ 16	IV	32
16	≤ 16	IV	32
17	≤ 16	III	32
18	≤ 16	III	32
$\downarrow$	$\downarrow$	$\downarrow$	$\downarrow$
31	≤ 16	III	32

determined. Assume all three-variable and higher order interactions are insignificant.

- Resolution III: all main variable effects can be determined. Assume all interactions are insignificant.

Then use Table 19.9 to select the appropriate matrix given the number of variables, sample size, and resolution. The matrix determines how to process the samples.

Now that we know the procedure to design experiments, details of the Hadamard matrix along with example experiments will be discussed.

4.2.1. Order 8 Hadamard Matrix

The order 8 Hadamard matrix is shown in Table 19.10. It consists of 8 rows and 7 contrast columns. The rows correspond to trials or samples to be processed in the experiment. The columns correspond to variables used in the experiment.

4.2.1.1. Two Variables (Resolution V)

When there are two variables, the order 8 Hadamard matrix has the following column labels.

Trial	A 1	B 2	3	-AB 4	5	6	7
1	+	-	-	+	-	+	+
2	+	+	-	-	+	-	+
3	+	+	+	-	-	+	-
4	-	+	+	+	-	-	+
5	+	-	+	+	+	-	-
6	-	+	-	+	+	+	-
7	-	-	+	-	+	+	+
8	-	-	-	-	-	-	-

For each trial, produce the samples as designated in columns 1 and 2. For instance, trial 1 is processed with variable A at the high (+) level and variable B at the low (−) level. Trial 2 is processed with A and B at the high (+) level and so on.

Trial	1	2	3	4	5	6	7
		Table 19.10. Order 8 Hadamard Matrix					
1	+	-	-	+	-	+	+
2	+	+	-	-	+	-	+
3	+	+	+	-	-	+	-
4	-	+	+	+	-	-	+
5	+	-	+	+	+	-	-
6	-	+	-	+	+	+	-
7	-	-	+	-	+	+	+
8	-	-	-	-	-	-	-

To estimate the effect of the main variables and the AB interaction, go down the column multiplying the result by +1 for a (+) and -1 for a (−). Add the results of the column, and divide by four since there are four pairs estimating the effect for an order 8 matrix.

$$\left(\overline{X}_{\text{high A}} - \overline{X}_{\text{low A}}\right) = \frac{(T1 + T2 + T3 - T4 + T5 - T6 - T7 - T8)}{4}$$

$$\left(\overline{X}_{\text{high B}} - \overline{X}_{\text{low B}}\right) = \frac{(-T1 + T2 + T3 + T4 - T5 + T6 - T7 - T8)}{4}$$

$$\left(\overline{X}_{\text{high AB}} - \overline{X}_{\text{low AB}}\right) = \frac{(T1 - T2 - T3 + T4 + T5 + T6 - T7 - T8)}{4}$$

This design is a resolution V design because the AB interaction can be estimated separately.

There are four unlabeled contrast columns—columns 3, 5, 6, and 7. Each unlabeled column can be used to estimate variance with one degree of freedom. Since there are four unlabeled columns, the variance can be estimated with four degrees of freedom. Estimate variance for column i using the following formula:

$$S_i^2 = \frac{\left[\sum (\text{column sign})(\text{result})\right]^2}{T}, \text{ with one degree of freedom} \quad (19.3)$$

where T is the number of trials. For instance, to estimate variance from column 3, use Equation (19.3) as follows.

$$S_3^2 = \frac{(-T1 - T2 + T3 + T4 + T5 - T6 + T7 - T8)^2}{8}$$

Once the variance estimates have been calculated for each of the unlabeled columns, obtain the average estimate.

$$S_{avg}^2 = \frac{\sum_i S_i^2}{j}, \text{ with } j \text{ degrees of freedom} \qquad (19.4)$$

where i is the column number and j is the number of columns used in the variance estimate.

Resolution V Matrix Design Example (2 variables)

During the development of a CMOS integrated circuit, it is characterized to determine under which conditions its yield on silicon wafers will be maximized. The engineer wants to determine the effects of varying the turn-on voltage of the NMOS and PMOS transistors and any interaction between the two. The response for this experiment is the yield of correctly functioning integrated circuits. Variance is unknown.

- **Objective**

  Define the variables and the levels for each.

Variable	-	+
A: NMOS turn-on voltage	0.6 V	0.8 V
B: PMOS turn-on voltage	-0.6 V	-0.8 V

  $(H_0)_1$: $\mu_{high\,A} = \mu_{low\,A}$

  $(H_a)_1$: $\mu_{high\,A} \neq \mu_{low\,A}$

  $(H_0)_2$: $\mu_{high\,B} = \mu_{low\,B}$

  $(H_a)_2$: $\mu_{high\,B} \neq \mu_{low\,B}$

  $(H_0)_3$: $\mu_{AB\,interaction} = 0$

$(H_a)_3$: $|\mu_{AB\ interaction}| > 0$

The sign of the interaction is of no value to the experimenter.

- **Choose α, β, and δ**

  $\alpha = 0.1$

  $\beta = 0.1$

  $\delta = 2.5\sigma$

  σ^2 is unknown

- **Calculate sample size**

  First calculate the sample size based on the normal distribution.

  $$N = 2(U_\alpha + U_\beta)^2 \frac{\sigma^2}{\delta^2}$$

  $U_\alpha = 1.645$ (from Table 19.4, double-sided)

  $U_\beta = 1.282$ (from Table 19.3)

  $$N = 2(1.645 + 1.282)^2 \frac{\sigma^2}{(2.5\sigma)^2} = 2.74$$

  It is estimated that an order 8 Hadamard matrix can be used. For an order 8 Hadamard matrix, $N_{high} = N_{low} = 4$. There are 4 unlabeled columns that can be used to estimate the variance. Thus $\phi = 4$.

  Now recalculate the sample size based on the t distribution.

  $$N_{high} = N_{low} = 2(t_\alpha + t_\beta)^2 \frac{\sigma^2}{\delta^2}$$

  $t_\alpha = 2.13$ (from Table 19.6, double-sided)

  $t_\beta = 1.53$ (from Table 19.5)

  $$N_{high} = N_{low} = 2(2.13 + 1.53)^2 \frac{\sigma^2}{(2.5\sigma)^2} = 4.29$$

  Round off to 4.

Use Table 19.9 to determine the matrix for the experiment. The required sample size of 4 with two variables can be accommodated by an order 8 Hadamard matrix. If the required sample size had been larger, a larger matrix would have to be used.

- **Define trials**

  Label the order 8 Hadamard matrix.

Trial	A 1	B 2	3	-AB 4	5	6	7
1	+	-	-	+	-	+	+
2	+	+	-	-	+	-	+
3	+	+	+	-	-	+	-
4	-	+	+	+	-	-	+
5	+	-	+	+	+	-	-
6	-	+	-	+	+	+	-
7	-	-	+	-	+	+	+
8	-	-	-	-	-	-	-

- **Perform the experiment**

  Determine the trial combinations from the matrix, conduct the trials, and test for yield.

Trial	A NMOS, V	B PMOS, V	Yield (%)
1	0.8	-0.6	65
2	0.8	-0.8	40
3	0.8	-0.8	44
4	0.6	-0.8	77
5	0.8	-0.6	68
6	0.6	-0.8	72
7	0.6	-0.6	78
8	0.6	-0.6	85

- **Multiply the test results by the response vectors and sum the columns**

Trial	A 1	B 2	3	-AB 4	5	6	7
1	+65	-65	-65	+65	-65	+65	+65
2	+40	+40	-40	-40	+40	-40	+40
3	+44	+44	+44	-44	-44	+44	-44
4	-77	+77	+77	+77	-77	-77	+77
5	+68	-68	+68	+68	+68	-68	-68
6	-72	+72	-72	+72	+72	+72	-72
7	-78	-78	+78	-78	+78	+78	+78
8	-85	-85	-85	-85	-85	-85	-85
Total	-95	-63	+5	+35	-13	-11	-9

- **Calculate the effects of the variables and interaction**

$$\overline{X}_{\text{high A}} - \overline{X}_{\text{low A}} = \frac{-95}{4} = -23.75$$

$$\overline{X}_{\text{high B}} - \overline{X}_{\text{low B}} = \frac{-63}{4} = -15.75$$

$$\overline{X}_{\text{high AB}} - \overline{X}_{\text{low AB}} = \frac{35}{4} = 8.75$$

- **Estimate the variance by using the results from the unlabeled columns**

  Use Equation (19.3) to estimate the variance.

$$\text{From column 3: } S_3^2 = \frac{(5)^2}{8} = 3.125$$

$$\text{From column 5: } S_5^2 = \frac{(-13)^2}{8} = 21.125$$

$$\text{From column 6: } S_6^2 = \frac{(-11)^2}{8} = 15.125$$

$$\text{From column 7: } S_7^2 = \frac{(-9)^2}{8} = 10.125$$

  Obtain an average estimate of variance using Equation (19.4).

$$S_{\text{avg}}^2 = \frac{3.125 + 21.125 + 15.125 + 10.125}{4} = 12.375 \text{ with } \phi = 4$$

$$S = 3.52$$

- **Calculate criterion**

$$\left| \overline{X}_{\text{high}} - \overline{X}_{\text{low}} \right|^* = t_\alpha S \sqrt{\frac{1}{N_{\text{high}}} + \frac{1}{N_{\text{low}}}}$$

  For the double-sided alternative hypotheses:

$t_\alpha = 2.13$ (from Table 19.6)

$$\left|\overline{X}_{high} - \overline{X}_{low}\right|^* = (2.13)(3.52)\sqrt{\frac{1}{4} + \frac{1}{4}} = 5.3$$

For the single-sided alternative hypothesis:

$t_\alpha = 1.53$ (from Table 19.5)

$$\left|\overline{X}_{high} - \overline{X}_{low}\right|^* = (1.53)(3.52)\sqrt{\frac{1}{4} + \frac{1}{4}} = 3.8$$

- **Compare results with the criterion**

  1. $\left(\overline{X}_{low\ A} - \overline{X}_{high\ A} = 23.75\right) > \left(\left|\overline{X}_{high} - \overline{X}_{low}\right|^* = 5.3\right)$

     Accept $(H_a)_1$: $\mu_{low\ A} > \mu_{high\ A}$ with at least 95% confidence (H_a is double-sided).

  2. $\left(\overline{X}_{low\ B} - \overline{X}_{high\ B} = 15.75\right) > \left(\left|\overline{X}_{high} - \overline{X}_{low}\right|^* = 5.3\right)$

     Accept $(H_a)_2$: $\mu_{low\ B} > \mu_{high\ B}$ with at least 95% confidence (H_a is double-sided).

  3. $\left(\overline{X}_{high\ AB} - \overline{X}_{low\ AB} = 8.75\right) > \left(\left|\overline{X}_{high} - \overline{X}_{low}\right|^* = 3.8\right)$

     Accept $(H_a)_3$: $\mu_{AB} > 0$ with at least 90% confidence (H_a is single-sided).

- **Determine nature of interaction**

  Since the AB interaction is significant, determine the nature of the interaction by plotting.

	Low B (PMOS = -0.6)	High B (PMOS = -0.8)
Low A (NMOS = 0.6)	78 85 $\overline{X} = 81.5$	77 72 $\overline{X} = 74.5$
High A (NMOS = 0.8)	65 68 $\overline{X} = 66.5$	40 44 $\overline{X} = 42$

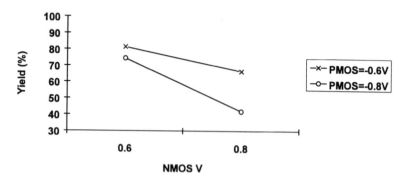

- **Make decision**

    Changing the NMOS turn-on voltage from 0.6 to 0.8 decreases yield by 15% when the PMOS turn-on voltage is -0.6, and it decreases yield by 32.5% when the PMOS turn-on voltage is -0.8. Changing the PMOS turn-on voltage from -0.6 to -0.8 decreases yield by 7% when the NMOS turn-on voltage is 0.6, and it decreases yield by 24.5% when the NMOS turn-on voltage is 0.8. The best yield is obtained by a combination of the NMOS turn-on voltage at 0.6 and the PMOS turn-on voltage at -0.6.

4.2.1.2. Three Variables (Resolution V)

Table 19.9 shows an order 8 Hadamard matrix can be used to design a resolution V experiment when there are three variables or less and $N_{high} = N_{low}$ is four or less. For three variables, label the columns as follows.

Trial	A 1	B 2	C 3	-AB 4	-BC 5	ABC 6	-AC 7
1	+	-	-	+	-	+	+
2	+	+	-	-	+	-	+
3	+	+	+	-	-	+	-
4	-	+	+	+	-	-	+
5	+	-	+	+	+	-	-
6	-	+	-	+	+	+	-
7	-	-	+	-	+	+	+
8	-	-	-	-	-	-	-

This design is similar to the full factorial design with three variables presented earlier. However with this design, there is a way to estimate the interactions. All main variables and all possible combinations of interactions are covered with separate contrast columns. This design is Resolution V because all two-variable interactions can be estimated separately.

Also note all the contrast columns are being used. There are no columns left to estimate variance. So to design an experiment in this way, the engineer must already know σ or S. If the engineer does not know σ or S, there are three options to consider.

1) The experiment can be replicated. Thus there are 16 trials with 8 combinations instead of 8 trials with 8 combinations. Each combination is performed twice. An estimate of variance can be obtained for each combination with one degree of freedom by using Equation (16.2). An estimate of variance with 8 degrees of freedom is obtained by repeating the experiment.

2) If it is known or strongly suspected certain interactions are not possible, the columns for these interactions can be used to estimate variance. For instance a three-variable interaction (ABC) is typically highly unlikely. Thus in many cases, column 6 can be used to estimate variance with one degree of freedom.

3) If the experiment variables are continuous in nature, such as temperature, time, and voltage, a number of samples can be run at the center point. These center point samples are run in addition to the trials which are defined by the matrix. The center point is the point halfway between the low and high levels of each variable. These center point samples can then be used to estimate variance.

4.2.1.3. Four Variables (Resolution IV)

To obtain a resolution IV design with an order 8 Hadamard matrix for four variables, label the columns as follows.

Trial	A 1	B 2	C 3	-CD -AB 4	-AD -BC 5	D ABC 6	-BD -AC 7
1	+	-	-	+	-	+	+
2	+	+	-	-	+	-	+
3	+	+	+	-	-	+	-
4	-	+	+	+	-	-	+
5	+	-	+	+	+	-	-
6	-	+	-	+	+	+	-
7	-	-	+	-	+	+	+
8	-	-	-	-	-	-	-

Note variable D is placed in column 6. This column had previously been used for the three-variable ABC interaction. To achieve a resolution IV design, it must be assumed all three-variable and higher order interactions are insignificant. This allows using column 6 for variable D.

Also note every possible combination of two-variable interactions are listed in the column headings. However instead of having a column all to itself, the two-variable interactions are confounded with another two-variable interaction. Thus if column 4 proves to be significant, it may not be possible to distinguish whether the AB interaction, the CD interaction, or both are the significant interactions. A second experiment with higher resolution needs to be performed to resolve the statistically significant interactions.

The confounding of the two-variable interactions is the nature of a resolution IV design. It is up to the engineer to pick the resolution while planning the experiment. If it is necessary to distinguish all two-variable interactions separately, a resolution V design is required. Table 19.9 shows a resolution V design with four variables can be obtained with an order 16 Hadamard matrix. However there may be constraints due to cost, time, or materials which limit the number of trials to eight. In these instances, the resolution IV design can be a very useful tool.

Resolution IV Matrix Design Example (4 variables)

This example revisits the problem that was identified in the $\overline{X}$, S control chart example (see Chapter 18, Section 3.3). DOE will be used to solve this problem. Recall that during the production of an integrated circuit, one of the steps involves depositing an insulator layer. A control chart is being used to monitor the insulator thickness, and this chart shows that the process is out of control. The target thickness is 100 nm, and the upper control limit for the distribution of $\overline{X}$ values is 106.2 nm. Some of the $\overline{X}$ values are above the upper control limit. The estimate of standard deviation (S) is known from the control chart and is equal to 5.0 nm. This estimate of standard deviation is based on 20 subgroups and each subgroup has 14 degrees of freedom ($\phi = N - 1 = 14$). Thus this estimate of standard deviation is based on 280 degrees of freedom. Since $\phi > 60$, this estimate is equal to the population standard deviation (σ) for all practical purposes. Set the significant difference (δ) at 3σ or 15 nm.

It is believed there are four variables needing control during deposition of the insulator: the temperature, power, pressure, and gas composition. Each of these will be examined over the range of their specified values.

- **Objective**

    To determine the likely cause for the insulator deposition process to be out of control. Define the suspected variables causing the thickness variation and the levels for each.

Variable	Specified Value	-	+
A: Temperature	$300 \pm 10°C$	290	310
B: Power	800 ± 50 W	750	850
C: Pressure	100 ± 10 mTorr	90	110
D: % Silane	$50 \pm 5\%$	45%	55%

$(H_0)_1$: $\mu_{high\,A} = \mu_{low\,A}$

$(H_a)_1$: $\mu_{high\,A} \neq \mu_{low\,A}$

$(H_0)_2$: $\mu_{high\,B} = \mu_{low\,B}$

$(H_a)_2$: $\mu_{high\,B} \neq \mu_{low\,B}$

$(H_0)_3$: $\mu_{high\,C} = \mu_{low\,C}$

$(H_a)_3$: $\mu_{high\,C} \neq \mu_{low\,C}$

$(H_0)_4$: $\mu_{high\,D} = \mu_{low\,D}$

$(H_a)_4$: $\mu_{high\,D} \neq \mu_{low\,D}$

$(H_0)_5$: $\mu_{AB+CD\ interaction} = 0$

$(H_a)_5$: $|\mu_{AB+CD \text{ interaction}}| > 0$

$(H_0)_6$: $\mu_{AD+BC \text{ interaction}} = 0$

$(H_a)_6$: $|\mu_{AD+BC \text{ interaction}}| > 0$

$(H_0)_7$: $\mu_{BD+AC \text{ interaction}} = 0$

$(H_a)_7$: $|\mu_{BD+AC \text{ interaction}}| > 0$

The sign of the interactions is of no value to the experimenter.

- **Choose α, β, and δ**

  $\alpha = 0.02$

  $\beta = 0.05$

  $\sigma = 5.0$ nm

  $\delta = 3\sigma = 15$ nm

- **Calculate sample size**

  Calculate the sample size based on the normal distribution.

  $$N = 2\left(U_\alpha + U_\beta\right)^2 \frac{\sigma^2}{\delta^2}$$

  $U_\alpha = 2.326$ (from Table 19.4, double-sided)

  $U_\beta = 1.645$ (from Table 19.3)

  $$N = 2(2.326 + 1.645)^2 \frac{(5.0)^2}{(15)^2} = 3.5$$

  Table 19.9 shows an order 8 Hadamard matrix would be adequate since $N_{high} = N_{low} = 4$. This would be a resolution IV design meaning all main effects can be identified and two-variable interactions are confounded with one another.

- **Define trials**

  Label the order 8 Hadamard matrix.

Trial	A 1	B 2	C 3	-CD -AB 4	-AD -BC 5	D 6	-BD -AC 7
1	+	-	-	+	-	+	+
2	+	+	-	-	+	-	+
3	+	+	+	-	-	+	-
4	-	+	+	+	-	-	+
5	+	-	+	+	+	-	-
6	-	+	-	+	+	+	-
7	-	-	+	-	+	+	+
8	-	-	-	-	-	-	-

- **Perform the experiment**

  Determine the trial combinations from the matrix, conduct the trials, and measure the insulator thicknesses.

Trial	A Temp.	B Power	C Pressure	D % Silane	Thickness (nm)
1	310	750	90	55	118.6
2	310	850	90	45	96.4
3	310	850	110	55	116.8
4	290	850	110	45	100.5
5	310	750	110	45	93.2
6	290	850	90	55	101.0
7	290	750	110	55	98.0
8	290	750	90	45	99.7

- **Multiply the test results by the response vectors and sum the columns**

Trial	A 1	B 2	C 3	-CD -AB 4	-AD -BC 5	D 6	-BD -AC 7
1	+118.6	-118.6	-118.6	+118.6	-118.6	+118.6	+118.6
2	+96.4	+96.4	-96.4	-96.4	+96.4	-96.4	+96.4
3	+116.8	+116.8	+116.8	-116.8	-116.8	+116.8	-116.8
4	-100.5	+100.5	+100.5	+100.5	-100.5	-100.5	+100.5
5	+93.2	-93.2	+93.2	+93.2	+93.2	-93.2	-93.2
6	-101.0	+101.0	-101.0	+101.0	+101.0	+101.0	-101.0
7	-98.0	-98.0	+98.0	-98.0	+98.0	+98.0	+98.0
8	-99.7	-99.7	-99.7	-99.7	-99.7	-99.7	-99.7
Total	+25.8	+5.2	-7.2	+2.4	-47.0	+44.6	+2.8

- **Calculate the effects of the variables and interactions**

$$\overline{X}_{high\ A} - \overline{X}_{low\ A} = \frac{25.8}{4} = 6.45$$

$$\overline{X}_{\text{high B}} - \overline{X}_{\text{low B}} = \frac{5.2}{4} = 1.3$$

$$\overline{X}_{\text{high C}} - \overline{X}_{\text{low C}} = \frac{-7.2}{4} = -1.8$$

$$\overline{X}_{\text{high D}} - \overline{X}_{\text{low D}} = \frac{44.6}{4} = 11.15$$

$$\left| \overline{X}_{\text{high AB+CD}} - \overline{X}_{\text{low AB+CD}} \right| = \frac{2.4}{4} = 0.6$$

$$\left| \overline{X}_{\text{high AD+BC}} - \overline{X}_{\text{low AD+BC}} \right| = \frac{47.0}{4} = 11.75$$

$$\left| \overline{X}_{\text{high AC+BD}} - \overline{X}_{\text{low AC+BD}} \right| = \frac{2.8}{4} = 0.7$$

- **Calculate criterion**

$$\left| \overline{X}_{\text{high}} - \overline{X}_{\text{low}} \right|^{*} = U_{\alpha}\sigma\sqrt{\frac{1}{N_{\text{high}}} + \frac{1}{N_{\text{low}}}}$$

For the double-sided alternative hypotheses:

$U_{\alpha} = 2.326$ (from Table 19.4)

$$\left| \overline{X}_{\text{high}} - \overline{X}_{\text{low}} \right|^{*} = (2.326)(5.0)\sqrt{\frac{1}{4} + \frac{1}{4}} = 8.22$$

For the single-sided alternative hypotheses:

$U_{\alpha} = 2.054$ (from Table 19.3)

$$\left| \overline{X}_{\text{high}} - \overline{X}_{\text{low}} \right|^{*} = (2.054)(5.0)\sqrt{\frac{1}{4} + \frac{1}{4}} = 7.26$$

- **Compare results with the criterion**

   Accept $(H_a)_4$: $\mu_{\text{high D}} > \mu_{\text{low D}}$ with a 99% confidence.

Accept $(H_a)_6$: $\mu_{AD+BC} > 0$. There is an AD and/or BC interaction with a 98% confidence.

Accept H_0 for all other hypotheses with a 95% confidence.

- **Determine nature of interaction**

The nature of resolution IV designs is that two-variable interactions are confounded with other two-variable interactions. Thus it is not possible to prove statistically which of the interactions (or both) are significant when a resolution IV, two-variable interaction column is significant. The following is a way to make an educated guess. This needs to be followed by an experiment of resolution V or higher to resolve the two-variable interactions statistically.

Make up two-variable tables to examine the AD and BC interactions.

AD Interaction

	Low D	High D
Low A	100.5	101.0
	99.7	98.0
	$\overline{X}=100.1$	$\overline{X}=99.5$
High A	96.4	118.6
	93.2	116.8
	$\overline{X}=94.8$	$\overline{X}=117.7$

BC Interaction

	Low C	High C
Low B	118.6	93.2
	99.7	98.0
	$\overline{X}=104.15$	$\overline{X}=95.6$
High B	96.4	116.8
	101.1	100.5
	$\overline{X}=98.7$	$\overline{X}=108.65$

- The AD interaction table is consistent with the findings that D is significant only when A is high. This points out there is also an AD interaction.

- The two highest measurements of thickness of the 8 trials occur when both A and D are high. This again points to an AD interaction.

- The BC interaction table shows some inconsistencies. There is a wide disparity of measurements for the case of high B and C and the case of low B and C.

The evidence indicates an AD interaction exists, and a BC interaction does not exist.

- **Make decision**

This experiment has shown it is possible to get significantly greater insulator thicknesses while staying within the specified values of the controllable parameters. The main effect causing the increase in thickness is an increase in % silane. There is also evidence an interaction between the temperature and % silane exists. Both of these parameters need to be controlled more closely in order to bring the insulator thickness back under control.

4.2.1.4. Five, Six, or Seven Variables (Resolution III)

Resolution III designs can estimate only main variable effects since all interactions are assumed to be insignificant. These designs can be very useful in the early stage of product or process development. The experimental space may be largely unknown in development engineering. A first experiment can help define the boundaries of the experimental space by including as many variables as possible. This first experiment is usually a resolution III design. A second experiment of higher resolution such as resolution V should be performed to confirm the results of the first experiment in more detail. This second experiment resolves whether main effects and/or two-variable interactions are significant.

As shown in Table 19.9, a resolution III design with an order 8 Hadamard matrix can be achieved with five, six, or seven variables. With seven variables, label the contrast columns as follows.

Trial	A 1	B 2	C 3	D 4	E 5	G 6	F 7
1	+	-	-	+	-	+	+
2	+	+	-	-	+	-	+
3	+	+	+	-	-	+	-
4	-	+	+	+	-	-	+
5	+	-	+	+	+	-	-
6	-	+	-	+	+	+	-
7	-	-	+	-	+	+	+
8	-	-	-	-	-	-	-

With six variables, column 6 is not used to calculate main effects. With five variables, columns 6 and 7 are not used. These unused columns can be used to estimate the variance.

Resolution III Matrix Design Example (7 variables)

In developing a process to manufacture raw sugar from sugar canes, it is desirable to quantify the effects of seven variables.

Variable	-	+
A: Evaporator Temperature	110°C	130°C
B: Evaporator residence time	24 hr.	48 hr.
C: Crystallizer temperature	40°C	60°C
D: Crystallizer time	12 hr.	24 hr.
E: Centrifuge rotation rate	30 rpm	60 rpm
F: Centrifuge design	Vendor A	Vendor B
G: Centrifuge time	1 hr.	2 hr.

It is assumed all two-variable and higher order interactions are negligible. The process development objective is to maximize the sucrose yield. From previous work, it is estimated the standard deviation (S) is 1% ($= 1 \times 10^{-2}$) and the estimate of variance (S^2) is 1×10^{-4} with $\phi = 10$. The current process has a mean yield (μ) of 84.0%. A significant difference (δ) of 2.5% is desired.

- **Objective**

    $(H_0)_1: \mu_{high\ A} = \mu_{low\ A}$

    $(H_a)_1: \mu_{high\ A} \neq \mu_{low\ A}$

    $(H_0)_2: \mu_{high\ B} = \mu_{low\ B}$

    $(H_a)_2: \mu_{high\ B} \neq \mu_{low\ B}$

    These pairs of hypotheses continue through variable G.

- **Choose α, β, and δ and state S^2**

    $\alpha = 0.05$

    $\beta = 0.10$

    $\delta = 2.5\%$

    $S = 1.0\%$ $\qquad\qquad S^2 = 1 \times 10^{-4}$ with $\phi = 10$

- **Calculate sample size**

    Calculate the sample based on the t distribution.

$$N = 2\left(t_\alpha + t_\beta\right)^2 \frac{S^2}{\delta^2}$$

$t_\alpha = 2.23$ (from Table 19.6, double-sided)

$t_\beta = 1.37$ (from Table 19.5)

$$N = 2(2.23 + 1.37)^2 \frac{\left(1.0 \times 10^{-2}\right)^2}{\left(2.5 \times 10^{-2}\right)^2} = 4.1$$

Round off to 4.

An order 8 Hadamard matrix can be used with up to seven variables for a resolution III design (see Table 19.9).

- **Define trials**

  Label the order 8 Hadamard matrix.

Trial	A 1	B 2	C 3	D 4	E 5	G 6	F 7
1	+	-	-	+	-	+	+
2	+	+	-	-	+	-	+
3	+	+	+	-	-	+	-
4	-	+	+	+	-	-	+
5	+	-	+	+	+	-	-
6	-	+	-	+	+	+	-
7	-	-	+	-	+	+	+
8	-	-	-	-	-	-	-

- **Perform the experiment**

  Determine the trial combinations from the matrix, conduct the trials, and measure the sucrose yield.

Trial	A Evap. Temp.	B Evap. Time	C Crystal. Temp.	D Crystal. Time	E Centrifuge Rate	F Centrifuge Design	G Centrifuge Time	Sucrose Yield (%)
1	130	24	40	24	30.0	B	2.0	83.2
2	130	48	40	12	60.0	B	1.0	86.2
3	130	48	60	12	30.0	A	2.0	88.5
4	110	48	60	24	30.0	B	1.0	87.9
5	130	24	60	24	60.0	A	1.0	87.8
6	110	48	40	24	60.0	A	2.0	87.5
7	110	24	60	12	60.0	B	2.0	87.0
8	110	24	40	12	30.0	A	1.0	83.5

- **Multiply the test results by the response vectors and sum the columns**

Trial	A 1	B 2	C 3	D 4	E 5	G 6	F 7
1	+83.2	-83.2	-83.2	+83.2	-83.2	+83.2	+83.2
2	+86.2	+86.2	-86.2	-86.2	+86.2	-86.2	+86.2
3	+88.5	+88.5	+88.5	-88.5	-88.5	+88.5	-88.5
4	-87.9	+87.9	+87.9	+87.9	-87.9	-87.9	+87.9
5	+87.8	-87.8	+87.8	+87.8	+87.8	-87.8	-87.8
6	-87.5	+87.5	-87.5	+87.5	+87.5	+87.5	-87.5
7	-87.0	-87.0	+87.0	-87.0	+87.0	+87.0	+87.0
8	-83.5	-83.5	-83.5	-83.5	-83.5	-83.5	-83.5
Total	-0.2	+8.6	+10.8	+1.2	+5.4	+0.8	-3.0

- **Calculate the effects of the variables.**

$$\overline{X}_{high\ A} - \overline{X}_{low\ A} = \frac{-0.2}{4} = -0.05\%$$

$$\overline{X}_{high\ B} - \overline{X}_{low\ B} = \frac{8.6}{4} = 2.15\%$$

$$\overline{X}_{high\ C} - \overline{X}_{low\ C} = \frac{10.8}{4} = 2.7\%$$

$$\overline{X}_{high\ D} - \overline{X}_{low\ D} = \frac{1.2}{4} = 0.3\%$$

$$\overline{X}_{high\ E} - \overline{X}_{low\ E} = \frac{5.4}{4} = 1.35\%$$

$$\overline{X}_{high\ F} - \overline{X}_{low\ F} = \frac{-3.0}{4} = -0.75\%$$

$$\overline{X}_{high\ G} - \overline{X}_{low\ G} = \frac{0.8}{4} = 0.2\%$$

- **Calculate criterion**

$$\left| \overline{X}_{high} - \overline{X}_{low} \right|^* = t_\alpha S \sqrt{\frac{1}{N_{high}} + \frac{1}{N_{low}}} = (2.23)(1\%)\sqrt{\frac{1}{4} + \frac{1}{4}} = 1.58\%$$

- **Compare results with the criterion**

  Accept $(H_a)_2$: $\mu_{high\,B} > \mu_{low\,B}$ with a 97.5% confidence.

  Accept $(H_a)_3$: $\mu_{high\,C} > \mu_{low\,C}$ with a 97.5% confidence.

  Accept H_0 for all other hypotheses with a 90% confidence.

- **Make decision**

  Significant improvement in sucrose yield can be made by changing two variables: increasing the evaporator residence time (variable B) from 24 to 48 hours and increasing the crystallizer temperature (variable C) from 40°C to 60°C. The combined effects of the two variables is almost 5%. In trial 3 and trial 4 where both these variables were high, the results were 88.5% and 87.9%.

  The experimenter may want to improve this sucrose separation process further. A higher resolution experiment involving the evaporator residence time and crystallizer temperature should be performed to see if there are any interactions. The experimenter may also want to include the centrifuge rotation rate (variable E) since this variable had the next highest effect.

4.2.1.5. Constructing the Hadamard Matrix

This section covers how the order 8 Hadamard matrix is constructed. The same steps also apply for constructing larger Hadamard matrices.

An order 8 Hadamard matrix has $T = 8$ trials and can accommodate up to $T - 1 = 7$ variables. The primary vector forms the basis for constructing the matrix. For $T = 8$, the primary vector is:

+ + + − + − −

The steps for constructing the Hadamard matrix are as follows.

1) Put the primary vector in column format.

```
+
+
+
–
+
–
–
```

2) Construct the next column by taking the bottom sign, and make it the top sign of the next column. The rest of the signs shift down by one position. Repeat this until there are $T - 1$ columns.

3) Add a row of minus (–) signs at the bottom.

+	–	–	+	–	+	+
+	+	–	–	+	–	+
+	+	+	–	–	+	–
–	+	+	+	–	–	+
+	–	+	+	+	–	–
–	+	–	+	+	+	–
–	–	+	–	+	+	+
–	–	–	–	–	–	–

4) Number the columns and rows.

Trial	1	2	3	4	5	6	7
1	+	–	–	+	–	+	+
2	+	+	–	–	+	–	+
3	+	+	+	–	–	+	–
4	–	+	+	+	–	–	+
5	+	–	+	+	+	–	–
6	–	+	–	+	+	+	–
7	–	–	+	–	+	+	+
8	–	–	–	–	–	–	–

This table of an order 8 Hadamard matrix is the same as Table 19.10.

Table 19.11. Contrast Column Labels for
Order 8 Hadamard Matrices*

# Variables	Resolution	1	2	3	4	5	6	7
2	V	A	B		AB			
3	V	A	B	C	AB	BC		AC
4	IV	A	B	C	AB, CD	BC, AD	D	AC, BD
5	III	A	B	C	D	E		
6	III	A	B	C	D	E		F
7	III	A	B	C	D	E	G	F

* All the two-variable interactions in the table are actually minus interactions.

Table 19.11 summarizes how the contrast columns should be labeled depending on how many variables are used. Remember for resolution V and IV, all three-variable and higher order interactions are assumed to be insignificant. For resolution III, all interactions are assumed to be insignificant. Any unlabeled columns can be used to estimate variance.

4.2.2. Order 16 Hadamard Matrix

Order 16 Hadamard matrices specify $T = 16$ trials and can accommodate up to $T - 1 = 15$ variables. The steps for constructing the matrix are the same as described in the previous section. For $T = 16$, the primary vector is:

$$+ + + + - + - + + - - + - - -$$

Table 19.12 summarizes how the contrast columns are labeled depending on the number of variables.

4.2.3. Order 32 Hadamard Matrix

Order 32 Hadamard matrices specify $T = 32$ trials and can accommodate up to $T - 1 = 31$ variables. For $T = 32$, the primary vector is:

$$+ + + + + - - + + - + - - + - - - - + - + - + + + - + + - - -$$

Table 19.13 summarizes how the contrast columns are labeled depending on the number of variables.

Table 19.12. Contrast Column Labels for Order 16 Hadamard Matrices*

# Variables	Resolution	1	2	3	4	5	6	7	8	9	10	11	12	13	14	15
2	V	A	B													
3	V	A	B	C			BC			AC						
4	V	A	B	C	D		BC	CD		AC	BD					AD
5	V	A	B	C	D	AB	BC	CD	CE	AC	BD	DE	AE	E	BE	AD
6	IV	A	B	C	D	AB, DE+	BC, AF	CD, EF	E	AC, BF	BD, AE	F		CE, DF		AD, BE
7	IV	A	B	C	D	AB, DE+	BD, AF+	CD, EF+	E	AC, BF+	BD, AE+	F	G	CE, DF+		AD, BE+
8	IV	A	B	C	D	AB, DE+	BC, AF+	CD, EF+	E	AC, BF+	BD, AE+	F	G	CE, DF+	H	AD, BE+
9	III	A	B	C	D	I			E			F	G		H	
10	III	A	B	C	D	I	J		E			F	G		H	
11	III	A	B	C	D	I	J	K	E	L		F	G		H	
12	III	A	B	C	D	I	J	K	E	L		F	G		H	
13	III	A	B	C	D	I	J	K	E	L	M	F	G		H	
14	III	A	B	C	D	I	J	K	E	L	M	F	G	N	H	
15	III	A	B	C	D	I	J	K	E	L	M	F	G	N	H	O

* All two-variable interactions in the table are actually minus interactions.

Table 19.13. Contrast Column Labels for Order 32 Hadamard Matrices*

#Var.	Res.	1	2	3	4	5	6	7	8	9	10	11	12	13	14	15	16
2	V	A	B														
3	V	A	B	C			AC										
4	V	A	B	C	D		AC	BD									
5	V	A	B	C	D	E	AC	BD	CE								
6	V	A	B	C	D	E	AC	BD	CE			AE				BF	F
7	IV	A	B	C	D	E	AC, DF	BD, EG	CE	F	G	AE				EF	
8	IV	A	B	C	D	E	AC, DF+	BD, EG+	CE	F	G	AE	H			EF, GH	
9	IV	A	B	C	D	E	AC, DF+	BD, EG+	CE, GI	F	G	AE	H	I		EF, GH+	
10	IV	A	B	C	D	E	AC, DF+	BD, EG+	CE, GI+	F	G	AE, FJ	H	I	J	EF, GH+	
11	IV	A	B	C	D	E	AC, DF+	BD, EG+	CE, GI+	F	G	AE, FJ+	H	I	J	EF, GH+	K
12	IV	A	B	C	D	E	AC, DF+	BD, EG+	CE, GI+	F	G	AE, FJ+	H	I	J	EF, GH+	K
13	IV	A	B	C	D	E	AC, DF+	BD, EG+	CE, GI+	F	G	AE, FJ+	H	I	J	EF, GH+	K
14	IV	A	B	C	D	E	AC, DF+	BD, EG+	CE, GI+	F	G	AE, FJ+	H	I	J	EF, GH+	K
15	IV	A	B	C	D	E	AC, DF+	BD, EG+	CE, GI+	F	G	AE, FJ+	H	I	J	EF, GH+	K
16	IV	A	B	C	D	E	AC, DF+	BD, EG+	CE, GI+	F	G	AE, FJ+	H	I	J	EF, GH+	K
17	III	A	B	C	D	E	Q			F	G		H	I	J		K
18	III	A	B	C	D	E	Q	R		F	G		H	I	J		K
19	III	A	B	C	D	E	Q	R	S	F	G		H	I	J		K
20	III	A	B	C	D	E	Q	R	S	F	G		H	I	J		K
21	III	A	B	C	D	E	Q	R	S	F	G		H	I	J		K
22	III	A	B	C	D	E	Q	R	S	F	G	T	H	I	J	U	K
23	III	A	B	C	D	E	Q	R	S	F	G	T	H	I	J	U	K
24	III	A	B	C	D	E	Q	R	S	F	G	T	H	I	J	U	K
25	III	A	B	C	D	E	Q	R	S	F	G	T	H	I	J	U	K
26	III	A	B	C	D	E	Q	R	S	F	G	T	H	I	J	U	K
27	III	A	B	C	D	E	Q	R	S	F	G	T	H	I	J	U	K
28	III	A	B	C	D	E	Q	R	S	F	G	T	H	I	J	U	K
29	III	A	B	C	D	E	Q	R	S	F	G	T	H	I	J	U	K
30	III	A	B	C	D	E	Q	R	S	F	G	T	H	I	J	U	K
31	III	A	B	C	D	E	Q	R	S	F	G	T	H	I	J	U	K

Table 19.13. Contrast Column Labels for Order 32 Hadamard Matrices (continued)

#Var.	Res.	17	18	19	20	21	22	23	24	25	26	27	28	29	30	31
2	V															
3	V			AB	BC											
4	V			AB	BC	CD									AD	BE
5	V			AB	BC	CD	DE								AD	BE
6	V	CF		AB	BC	CD	DE		EF	AF		DF			AD	BE
7	IV	AG		AB	BC	CD, AF	DE, BG		BF	CG		FG			AD, CF	BE, DG
8	IV	AG		AB, CH	BC, AH	CD, AF	DE, BG		BF, DH	CG		FG, EH			AD, CF	BE, DG
9	IV	AG		AB, CH+	BC, AH+	CD, AF+	DE, BG		BF, DH+	CG, EI		FG, EH			AD, CF+	BE, DG
10	IV	AG, HJ		AB, CH+	BC, AH+	CD, AF+	DE, BG+		BF, DH+	CG, EI+		FG, EH			AD, CF+	BE, DG+
11	IV	AG, HJ+		AB, CH+	BC, AH+	CD, AF+	DE, BG+		BF, DH+	CG, EI+		FG, EH+			AD, CF+	BE, DG+
12	IV	AG, HJ+	L	AB, CH+	BC, AH+	CD, AF+	DE, BG+		BF, DH+	CG, EI+		FG, EH+			AD, CF+	BE, DG+
13	IV	AG, HJ+	L	AB, CH+	BC, AH+	CD, AF+	DE, BG+	M	BF, DH+	CG, EI+		FG, EH+			AD, CF+	BE, DG+
14	IV	AG, HJ+	L	AB, CH+	BC, AH+	CD, AF+	DE, BG+	M	BF, DH+	CG, EI+	N	FG, EH+			AD, CF+	BE, DG+
15	IV	AG, HJ+	L	AB, CH+	BC, AH+	CD, AF+	DE, BG+	M	BF, DH+	CG, EI+	N	FG, EH+	O		AD, CF+	BE, DG+
16	IV	AG, HJ+	L	AB, CH+	BC, AH+	CD, AF+	DE, BG+	M	BF, DH+	CG, EI+	N	FG, EH+	O	P	AD, CF+	BE, DG+
17	III		L					M			N		O	P		
18	III		L					M			N		O	P		
19	III		L					M			N		O	P		
20	III		L					M			N		O	P		
21	III		L					M			N		O	P		
22	III	V	L	W				M			N		O	P		
23	III	V	L	W	X			M			N		O	P		
24	III	V	L	W	X	Y		M			N		O	P		
25	III	V	L	W	X	Y	Z	M			N		O	P		
26	III	V	L	W	X	Y	Z	M			N		O	P		
27	III	V	L	W	X	Y	Z	M	A1		N		O	P		
28	III	V	L	W	X	Y	Z	M	A1	B1	N		O	P		
29	III	V	L	W	X	Y	Z	M	A1	B1	N	C1	O	P		
30	III	V	L	W	X	Y	Z	M	A1	B1	N	C1	O	P	D1	
31	III	V	L	W	X	Y	Z	M	A1	B1	N	C1	O	P	D1	E1

* All two-variable interactions in the table are actually minus interactions.

KEY POINTS: Statistics

1) Statistics provides a set of tools for you to approximate the true state of nature. This is done by observing the central tendency and distribution of samples taken from the population. This information can be used to distinguish true signals (either desirable or undesirable) versus noise. Identify and focus on the signals to be most effective at advancing research projects, improving the product development cycle, and increasing manufacturing efficiency.

2) Use Statistical Process Control (SPC) to identify problems in a production environment. Development engineers need to consider process capability when setting specifications for new products and processes.

3) Use Design of Experiments (DOE) to solve problems either in a research and development or production environment.

4) Take the time to plan carefully any use of SPC or DOE. Control charts need to examine the most important parameters, consist of rational subgroups, and be practical in order to be successful. Define the objective of any experiment as specifically as possible and consider all variables in order to reach correct decisions based on the data.

5) For SPC, use $\overline{X}$, R and $\overline{X}$, S control charts for variable data. Use p control charts for attribute data.

6) For DOE, use matrix designs for multi-variable experiments. They are superior to the one-variable-at-a-time method. Select the resolution of the experiment based on the amount of interaction information desired, time, and costs. For single-variable experiments, use simple comparative methods.

7) Use statistics to make decisions and take immediate action. Once a significant change has been identified by SPC, find the cause. If using DOE determines a new product or process is feasible, complete the project as quickly as possible. On the other hand if using DOE determines the project is not feasible, kill it. In this way, using statistics achieves the maximum benefit for the minimum cost.

SUGGESTED ACTIVITIES: Statistics

1) Think of an application for the use of control charts, preferably in your field, involving a process or an equipment characteristic. Determine the most applicable control charts by examining Table 18.5. Follow the steps outlined in Table 18.1 to plan the charts, and gather and analyze the data.

2) Calculate the process capability for the example in Activity (1) by using either pre-determined or arbitrary specification limits. Estimate the fraction of the population falling outside the specification limits by using Table 18.4.

3) Construct a chart for Pareto analysis in order to focus your problem solving efforts. This type of chart can also be used to analyze how you spend your time as discussed in Part I under Time Management.

4) Think of a specific problem of comparing two means (such as two methods or brands). Determine which case you have as described in the "Single Variable Experiments—Comparing Means" section (Chapter 19, Section 3). Then perform the steps outlined in Table 19.1 to plan and perform your experiment and make the decision.

5) Construct an order 16 Hadamard matrix by using the primary vector shown in the "Order 16 Hadamard Matrix" section (Chapter 19, Section 4.2.2.).

6) Design a two-level, multi-variable experiment concerning a process or product you are trying to develop, understand, or troubleshoot. If an engineering application doesn't come immediately to mind, optimize a process which affects your everyday life such as the time it takes to commute to work or school. Follow the procedure described in the "Matrix Designs" section (Chapter 19, Section 4.2). Plan and perform the experiment, and make your decisions based on the results.

REFERENCES: Statistics

Blank, L., *Statistical Procedures for Engineering, Management and Science*, New York: McGraw-Hill, 1980.

Bowker, A. H., and Goode, H. P., and Lieberman, G. J., *Engineering Statistics, 2nd Edition*, Englewood Cliffs, N.J.: Prentice-Hall, 1972.

Box, G. E. P., and Draper, N. R., *Empirical Model-Building and Response Surfaces*, New York: John Wiley & Sons, 1986.

Box, G. E. P., and Hunter, W. G., and Hunter, J .S., *Statistics for Experimenters*, New York: John Wiley & Sons, 1978.

Cochran, W. G., and Cox, G. M., *Experimental Designs, 2nd Edition*, New York: John Wiley & Sons, 1957.

Cox, D. R., *Planning of Experiments*, New York: John Wiley & Sons, 1958.

Daniel, C., *Applications of Statistics to Industrial Experimentation*, New York: John Wiley & Sons, 1976.

Deming, W. E., *Quality, Productivity, and Competitive Position*, Cambridge, Mass.: The MIT Press, 1982.

Diamond, W. J., *Practical Experiment Designs for Engineers and Scientists, 2nd Edition*, New York: Van Nostrand Reinhold, 1989.

Grant, E. L. and Leavenworth, R. S., *Statistical Quality Control, 6th Edition*, New York: McGraw-Hill, 1988.

Guttman, I., and Wilks, S. S., *Introductory Engineering Statistics, 3rd Edition*, New York: John Wiley & Sons, 1982.

Ishikawa, K., *Guide to Quality Control, 2nd Edition*, Tokyo: Asian Productivity Organization, 1982.

John, P. W. M., *Statistical Design and Analysis of Experiments*, New York: The Macmillan Company, 1971.

Juran, J. M. (ed.), *Quality Control Handbook, 3rd Edition*, New York: McGraw-Hill, 1974.

Mendenhall, W., *Introduction to Probability and Statistics, 6th Edition*, N. Scituate, Mass.: Duxbury Press Divison of Wadsworth Publishing Co., 1983.

Natrella, M. G., *Experimental Statistics*, National Bureau of Standards Handbook 91, Washington D.C.: U.S. Government Printing Office, 1966.

Scheaffer, R. L., and McClave, J. T., *Statistics for Engineers*, Boston: Duxbury Press, 1982.

Shewhart, W. A., *Economic Control of Quality of Manufactured Product*, Princeton, N.J.: Van Nostrand Company, 1931. Reprinted by American Society for Quality Control, Milwaukee, Wis.

Taguchi, G., *Introduction to Quality Engineering: Designing Quality into Products and Processes*, White Plains, N.Y.: Kraus International Publications, 1986.

Volk, W. *Applied Statistics for Engineers, 2nd Edition*, New York: McGraw-Hill, 1969.

Western Electric Company, *Statistical Quality Control Handbook, 2nd Edition*, New York: Western Electric Company, 1958.

PART V

PROJECT MANAGEMENT

Putting It All Together

With the business environment becoming increasingly competitive, product life cycles are becoming shorter. This means there is emphasis on new product introductions with short product development times. There is also increasing pressure to produce these products as efficiently as possible while maintaining quality objectives. Project management helps to better use existing company resources to develop products and services more effectively and to produce them more efficiently.

All engineers are called upon to manage projects of various kinds. Even in Engineering School, students are involved with projects. The scope and duration of projects assigned to you depend on your level of experience. Project durations can be

- short (within weeks)
- medium (within months)
- long (greater than six months).

For instance, engineers with five years of experience or less may be assigned projects of short duration whereas engineering managers with 15 to 20 years of experience may be assigned projects of long duration. The financial impact of projects on the organization also depends on the scope and duration of the project.

However no matter the project's scope and duration, the concepts of managing projects are the same. Project management is the planning, implementing, and monitoring of the organization's resources to achieve specific objectives. Engineers have a perfect background for project management because a technical background is necessary as well as an understanding of how technology can be applied.

In order to manage projects successfully, you must understand and use the concepts presented in Parts I through IV as well as new concepts presented in Part V.

Here are the keys to successful project management.

1) **Communicate effectively**
   Part I covered the importance of good communication skills when communicating with yourself and other people. In order to be a successful project manager, utilize the skills described such as listening, getting your ideas across, getting

along with your boss, and getting along with your co-workers.

In addition, Part V on project management will describe other communication skills in order to be successful. Topics addressed include:

- negotiating to win support from others
- setting team roles and responsibilities
- running effective meetings
- resolving conflicts
- communicating with project status reports

2) Alignment with strategic plan

As stated in Part II, the purpose of a strategic plan is to define the most important tasks for the organization to achieve long-term success. As a project manager, see the big picture and understand how your project contributes towards achieving tasks in the strategic plan. Continually communicate this link to the executives, functional managers, and project team members. Showing the importance of your project is necessary to obtain support and commitment from these people.

3) Contribute to key financial measures

Part III showed you must understand how you can make a difference. Part of this is knowing the strategic plan. Another part is knowing how the financial plan fits in with the strategic plan and knowing the key financial measures for the organization. See how your project contributes toward improving key financial measures. Communicate this link to those involved with the project.

Part V will take the subject of finance one step further by showing how projects can be evaluated in terms of their financial, strategic, and technical merits. If a link between your project and the organization's strategic and financial plans cannot be identified, question the value of your project. Perhaps there are other project ideas that would contribute more significantly to your organization's strategic and financial plans.

4) Use statistics whenever possible

Part IV showed statistics can be used to improve confidence levels when making technical decisions. The tools of Statistical Process Control and Design of Experiments return the maximum amount of information for the minimum cost. Use statistics for projects both in development and production.

5) Utilize project management tools during each phase of the project

The five phases of managing a project are evaluation, planning, implementation, monitoring, and closing. Part V will present specific tools to use in each of these phases. The evaluation chapter will cover the basics of two financial methods: net present value and return on investment. The planning chapter will show the elements of a sound plan and how to win support from others. The chapter on implementation describes traits of a successful project manager and includes pointers on how to run meetings and how to resolve conflicts. The chapter on monitoring the project will offer ways to communicate project progress. Finally the closing chapter will show how to properly end your project.

The key to successful project management is to tie together all the concepts presented in this book: communications, strategic planning, finance, statistics, and project management tools. Of course another key component is to understand the particular technology pertinent to the project. Now let's get started in going through the five phases of project management.

Chapter 20

PROJECT PHASES

A project is an endeavor using the organization's resources to achieve specific objectives. Project management is a way to effectively and efficiently use these resources in order to achieve the project objectives.

A project consists of five phases: evaluation, planning, implementation, monitoring, and closing. These are shown in Figure 20.1. There are several practical considerations to remember when managing a project.

- Evaluation: Every project needs to be initiated with the evaluation phase. Make sure you know how the project will be of benefit to your organization. In your own mind determine why the project is important. Make sure you do the right things before you dive into trying to do things right.

- Planning: The most important activity for a project manager is to plan. This phase is so important that it can take 50% of the entire project's resources. Sound planning takes more time up front before the implementation phase begins. But in the long run, planning will save time in the implementation, monitoring, and closing phases. A sound plan anticipates problems and determines solutions before they occur. A sound plan requires commitment from the appropriate managers and all project team members.

- Planning/Implementation/Monitoring: These three phases are iterative. Even with sound planning, there will be situations not planned for. Planning is a continual process even during project implementation.

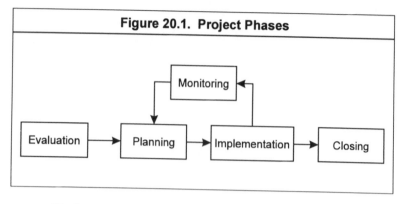

Figure 20.1. Project Phases

- <u>Closing</u>: Successful projects have a well-defined closing phase. The specifics of the project closing should be accounted for in the planning phase. Planning for the end of the project includes how the project team members will be transferred to other projects, how information and technology will be transferred, and how achievements will be awarded.

An effective project manager guides the project through these phases and utilizes appropriate tools in each phase. These phases and tools will now be covered in more detail.

Chapter 21

EVALUATING PROJECTS

The first phase of any project is the evaluation phase. The purpose of this phase is to analyze the benefits and costs of the project versus other project ideas. Projects are evaluated by the executives. Projects that meet predetermined cost/benefit criteria will be selected by the executives to proceed to the planning phase. The project manager may or may not be involved in the evaluation phase. As a project manager, if you get involved after the evaluation phase, you need to know why the project is being pursued and what are the overall benefits.

However even before the evaluation phase of a project can take place, the executives must establish a strategic plan and key financial measures in order to set a criteria for evaluating projects.

1. STRATEGIC ALIGNMENT

Referring back to Part II, one of the most important roles of the executives is to develop a strategic plan. Strategic planning examines three areas: the goals of the firm, the capabilities of the firm, and the business environment. The strategic plan defines a course of action that ties together these three areas to create and maintain a competitive advantage in order to create and maintain above average profitability.

The capabilities are the strengths and weaknesses of the firm's functional areas compared to other firms in the industry. Areas to consider when evaluating the firm's capabilities include the firm's technical base and manufacturing capability. The project under evaluation should play to your company's strengths and minimize areas where your company is weak. If the project requires expertise or resources where your company is weak, there needs to be a clear plan to develop the weakness into a strength. For example if the project is to design and build bridges and your firm's engineers have experience building only highways, there must be a sound plan to gain bridge expertise. Otherwise scrap the project. Play to your company's strengths.

Assess the business environment by analyzing five competitive factors:

- threat of new entrants
- threat of substitution
- bargaining power of buyers
- bargaining power of suppliers
- intensity of rivalry among competitors

The competitive factors in an industry determine the extent of potential profits for firms in the industry. Firms are most profitable under the following conditions. The threat of new entrants is low. An example is when firms have proprietary technical knowledge. The threat of substitution is low—there is little chance the product or service can be substituted with a new product or service with better price/performance characteristics. The bargaining power of the people buying the product or service is low, and the bargaining power of the suppliers to the industry is low. With low buyer bargaining power, prices for the product or service can be maintained or increased. Finally, another condition for continued profitability in an industry is when the intensity of rivalry among competitors is low.

A sound strategic plan considers the business environment and pursues operating in areas where the competitive factors are the weakest. Similarly a project which is part of such a strategic plan helps a firm get to where the competitive factors for an industry are weak.

A final area to consider when evaluating a project for strategic alignment is the goals of the firm. These goals include financial goals, attitude toward risk, and values. The more the project helps the firm achieve its strategic plan goals, the more important it is to the firm.

The bottom line is you must understand the firm's strategic plan as it defines the most important tasks to achieve long-term success. As a project manager, you need to see the link between a proposed project and how it helps the firm achieve its most important tasks. This link establishes priority for your project and provides motivation to successfully complete the project. Constantly communicate this link with the strategic plan to the executives, functional managers, and your project team. This helps establish and maintain their commitment to the project.

Strive to identify linkage with the strategic plan. If you cannot see how your project aligns with the strategic plan, question the value of your project. If the purpose of your project cannot be identified, recommend that other projects be evaluated and pursued. Strive to work on projects important to the firm. Unimportant projects waste company resources. Working on unimportant projects can be a dead end to your career advancement.

2. ENGINEERING ECONOMICS

Recall from Part III a business can be modeled as a dynamic system. There is the operations area and the investment area (see Figure 9.1). The operations area uses assets of the company to develop, manufacture, and market products or services to attain operating profits. The investment area determines how these profits are deployed and how much is reinvested into the operations area to fund projects for future growth and profitability.

The projects being funded in the operations area need to tie in with the company's strategic plan. These projects also need to show a sufficient return on investment (ROI) to justify the risk associated with investing in the project. This section describes how projects are evaluated in financial terms. Engineering economics will be used to analyze project cash flows in terms of net present value and return on investment.

2.1. Cash Flow

In order to analyze the financial side of the project, the project's cash flow must be estimated over time. Use historical data from similar projects whenever possible. Talk to the people who have the most direct knowledge of what the project's costs and benefits will be. For cost estimates, start with the functional managers in Engineering and Manufacturing. For benefits, talk with the Marketing personnel. Work with Accounting throughout this process to help guide you.

Once the project's cash flows have been estimated, graph them on a cash flow diagram to gain better understanding. Use the following conventions to graph a cash flow diagram.

- Plot time on the horizontal axis. Mark this axis in equal time periods.

- All cash flows are assumed to take place at the end of a period. The exception to this is the initial cost which takes place at $t=0$.

- When there is more than one cash flow activity (either receipts or disbursements) during the period, they may be combined.

Example 21.1: Cash Flow Diagram

An automated tester for a manufacturing line has a purchase price of $50,000. The automated testing capability will save the company $15,000 each year for five years. The tester maintenance costs are $2,000 per year. At the end of five years, it is estimated the tester can be sold for $10,000.

The cash flow diagram is plotted using one year periods. The first diagram shows each cash flow separately.

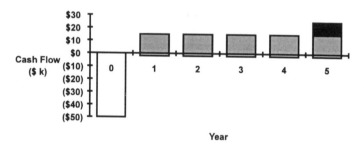

To simplify calculations, combine the cash flows in each period.

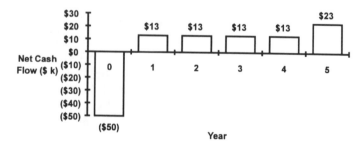

2.2. Equivalency

Engineering economics is concerned with the magnitude of the estimated cash flows as well as the timing of the cash flows. The concept of equivalency is used to compare alternative projects with different cash flows.

To illustrate the concept of equivalency, consider the case where you have $100. Suppose you can place this money in a savings account which pays 4% effective annual interest at the end of each year. At the end of the first year, the account will grow to $104. At the end of the second year, the account will grow to $108.16.

Assume you have no need for this money over the two years. Any money received during the two years will be put in the 4% savings account. Which alternative is more desirable?

- Alternative 1: $100 now

- Alternative 2: $104 at the end of one year

- Alternative 3: $108.16 at the end of two years

With the assumptions given, these three alternatives are equivalent. None of the alternatives is superior to the other two.

2.3. Computing Equivalency

In the previous section, it was shown $100 now is equivalent to $104 in one year given a savings account earning 4% interest. The $100 is known as the present value, P. The present value is the worth of a cash flow at time $t = 0$. The $104 is known as the future value, F, at time $t = 1$. In this case, the unit of time is in years. The future value is calculated by Equation (21.1).

$$F = P(1 + i)^n \qquad (21.1)$$

In Equation (21.1), i is the effective interest rate per period, and n is the number of periods. Equation (21.1) can be reconfigured to solve for the present value, P. This is shown in Equation (21.2).

$$P = F(1 + i)^{-n} \qquad (21.2)$$

Example 21.2: Future Value

How much will $100 be worth in five years if it is put into a 4% savings account?

Use Equation (21.1) with $P = \$100$, $i = 0.04$, and $n = 5$. The unit of n is in years.

$F = \$100(1 + 0.04)^5 = \$121.67$

Example 21.3: Present Value

How much do you need to put in a 6% savings account in order for it to be worth $10,000 in ten years?

Another way to look at this problem is to figure out the present value of $10,000 ten years from now assuming the interest rate is 6%. Use Equation (21.2) with $F = \$10,000$, $i = 0.06$, and $n = 10$. The unit of n is in years.

$P = \$10,000(1 + 0.06)^{-10} = \$5584$

2.4. Net Present Value (NPV)

The net present value is the sum of the present values of each component in the cash flow diagram. Consider the case where you deposit $100 now in a 4% savings account and you withdraw the $104 a year from now. The cash flow diagram looks like this.

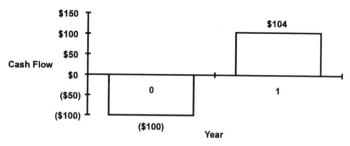

The net present value is:

$NPV = P_0 + P_1$ where P_n is the present value of the cash flow component for the period $t = n$.

$NPV = -\$100 + \$104(1 + 0.04)^{-1} = \$0$

The net present value in this example is $0. This means the alternatives of having $100 now and having $104 one year from now are equivalent given the conditions of this example.

Now suppose there is an alternative where you are offered $110 after one year for the use of your $100. The net present value at 4% for this case is:

$$NPV = P_0 + P_1$$
$$= -\$100 + \$110(1 + 0.04)^{-1}$$
$$= \$5.77$$

The net present value is positive and is a more desirable alternative than the previous case. Alternatives with a positive net present value are desirable because they increase the earnings of the money invested. Alternatives with a negative net present value are undesirable.

Example 21.4: Net Present Value

Calculate the net present value for the situation described in Example 21.1. Recall this is the example of purchasing an automated tester. The initial purchase price is $50,000. The automated testing capability will save the company $15,000 per year for five years. The maintenance costs are $2,000 per year, and it is estimated the tester can be sold for $10,000 in five years. Calculate the net present value at an effective interest rate of 8%. This interest rate has been determined by the company to be the minimum rate of return for any project under consideration.

To calculate the net present value, calculate the present value for each cash flow component and sum them up. Use the cash flow diagram shown in Example 21.1.

$$NPV = P_0 + P_1 + P_2 + P_3 + P_4 + P_5$$
$$= -\$50,000 + \$13,000(1 + 0.08)^{-1} + \$13,000(1 + 0.08)^{-2} +$$
$$\$13,000(1 + 0.08)^{-3} + \$13,000(1 + 0.08)^{-4} +$$
$$\$23,000(1 + 0.08)^{-5}$$
$$= -\$50,000 + \$12,037 + \$11,145 + \$10,320 + \$9,555 +$$
$$\$15,653$$
$$= \$8,711$$

The net present value at 8% of the project in Example 21.1 is a positive value. This means the return on investment is greater than 8%. This project is desirable.

A critical component in calculating net present value is selecting the interest rate to use. The interest rate should be the minimum acceptable rate of return for the projects under consideration. The minimum acceptable rate of return needs to be set by those people developing the strategic plan in conjunction with the Finance department. One way of setting the minimum rate of return is by calculating the average rate of return from past projects. A lower bound on the minimum acceptable rate of return is the rate that could be earned in a savings account or with a bond covering the duration of the project. In practice, the minimum acceptable rate of return will be higher than what can be earned in a bank because of the higher risks associated with investing in an engineering project.

A last note on net present value is this calculation is available in spreadsheets. Now that you know the theory behind net present value, you can set up your computer to do the calculating.

2.5. Return On Investment (ROI)

Another method for evaluating a project is by estimating the return on investment. The return on investment is the equivalent interest rate the money invested in the project would yield if the returns from a bank account were matched with cash flow returns from the project. The ROI is the interest rate which makes the net present value of a project equal zero.

The steps for calculating the return on investment are as follows:

1. Determine the cash flow diagram for the project.

2. Select a guess of the interest rate, i, and calculate the net present value.

3. If the net present value is positive, select a higher value of i. If the net present value is negative, select a lower value of i. Recalculate the net present value.

4. Repeat step 3 until the net present value is $0. The corresponding i is the return on investment.

An alternative way to calculate ROI is to use a computer spreadsheet. Once the cash flow by period is determined, the return on investment function can be used. The spreadsheet function may also be called internal rate of return (IRR) or rate of return (ROR).

Example 21.5: Return On Investment

For the project in Example 21.1, determine the return on investment.

Use the cash flow diagram shown in Example 21.1. Iteratively solve for i which corresponds to $NPV = \$0$.

i	NPV	Comment
8%	+$8711	NPV is positive, select higher i
16%	-$2673	NPV is negative, select lower i
14%	-$176	NPV is negative, select lower i
13.9%	-$46	NPV is negative, select lower i
13.8%	+$85	By rounding off, ROI=13.9%

3. SELECTING POTENTIAL PROJECTS

In order to select potential projects, they must be evaluated in terms of strategic alignment and engineering economics. This is the responsibility of the executives. The strategic alignment can be judged by how well the project fits in with the goals of the firm, the capabilities of the firm, and the business environment. The engineering economics can be quantified in terms of net present value or return on investment.

In addition, there are other factors to consider such as amount of risk and amount of investment. Once all the primary factors for evaluation of the projects have been determined, lay out all the projects under consideration in a table. Evaluate each project in terms of the primary factors. An example is shown in Table 21.1. In this table, the qualitative factors are evaluated in terms of low, medium, or high. The minimum acceptable return on investment is 8%.

Table 21.1. Evaluating Potential Projects

| Project | Strategic Alignment | | | ROI (minimum=8%) | Risk | Investment |
	Goals	Capability	Business Environment			
A	Medium	High	Medium	14%	Low	Medium
B	Low	Medium	High	5%	Medium	High
C	High	Low	High	25%	High	High
D	High	High	High	20%	Medium	Medium

Project	Comments
A	Low risk, medium to high strategic alignment
B	ROI below minimum acceptable
C	Low capability--need to train or hire workforce
D	High strategic alignment

The purpose of laying out the project in a table is to determine which projects, if any, to pursue. There may not be any clear cut winners, but the table facilitates the decision making process.

Let's evaluate Table 21.1 a little closer. Assume it is possible to invest in only one new project. Project B can be eliminated because the estimated return on investment is below the minimum acceptable rate. Project A has an ROI = 14% with only medium to high strategic alignment, but risk is low. Project D has a higher ROI than Project A with high strategic alignment, but risk is medium. And Project C has an even higher ROI than Project D, but risk is high and capability of the firm is low.

In selecting among Projects A, D, and C, the risk increases as the estimated ROI increases. Which project gets selected will be determined by the values of the firm. Is the firm a risk taker willing to go for an ROI = 25% even though risks are high? Or is it preferable to settle for an ROI = 14% with low risks? Or maybe the firm is comfortable with pursuing something in between.

Once a project has been selected and once you've been assigned as the project manager, you need to understand why the project was selected and how it fits in with the larger picture for the firm. This understanding is the critical first step for successful project management.

To summarize, the project has now gone through the evaluation phase. Executives have studied various project proposals in terms of strategic alignment, financial measures, and other factors. Hopefully you have been involved in this phase to ensure the estimates of time and cost are reasonable. Executives select projects they deem as desirable. They authorize taking the project to the next phase: project planning.

Chapter 22

PROJECT PLANNING

Project planning is the most important responsibility for a project manager. Developing a sound plan is the key difference between good and poor project management. Quality must be planned for and designed into the project from the outset. As much as half of the labor hours for the project could be spent on the evaluation and planning phases. The elements of a sound project plan are shown in Table 22.1. Once the elements of the project plan have been defined, executives need to approve the plan before going to the next phase of the project which is implementation.

Chapter 22 will describe each of the planning elements in more detail. This will be followed by methods to get the plan done by using certain communication skills.

1. PROJECT PLAN ELEMENTS

1.1. Project Objective

The first element of a project plan is defining the project objectives. These objectives should be summarized on a single page. Specifications for the project can be attached. An example form to show the project objectives is shown in Figure 22.1. This form is filled out with an example project in Figure 22.2.

The key to defining the project objectives is to be clear and concise. The language of the project objectives needs to be as unambiguous as possible. Emphasize results that can be measured.

In the Objectives section, state the major deliverable, when it will be delivered, and the target cost.

Table 22.1. Elements of a Project Plan
1. Project objectives
2. Work Breakdown Structure (WBS)
3. List of team members
4. Responsibility matrix
5. Schedule
6. Cost
7. Closure plan

In the Purpose section, summarize the background of why this project is important and desirable for the firm. Summarize here the reasons found during the evaluation phase of the project that led to its selection.

The Scope section is broken down into three areas: deliverables, measurables, and exclusions. The deliverables are the output items produced by the end of the project that will be used in the ongoing operations area of the firm. The measurables are standards used to evaluate the deliverables as criteria for acceptance. The measurables are typically specifications agreed upon by Engineering, Marketing, and the potential customer of the product or service. Exclusions are items that are not going to be done during the course of the project. Listing exclusions can be important to clear up any misunderstandings between Engineering, those receiving the project deliverables, and management.

1.2. Work Breakdown Structure

The next element of the project plan is the work breakdown structure (WBS). The WBS is a graphical representation of the tasks necessary to complete the project. It begins with the overall project, then breaks it down into smaller and smaller tasks. The levels that can be used in a WBS are:

Level 1:	Project
Level 2:	Major blocks
Level 3:	Tasks
Level 4:	Subtasks

Figure 22.1. Project Objectives Form

Project:
Project Manager:
Date:

Objectives:

Purpose:
    Strategy:

    Strategic Alignment:

    Financial Impact:

Scope:
    Deliverables:

    Measurables:

    Exclusions:

Figure 22.2. Project Objectives Example

Project: DVD-ROM Drive Controller
Project Manager: Lisa Johnson
Date: August 1, 1996

Objectives:
- To develop an integrated circuit controller for the next generation DVD-ROM drives.
- To make the first production units available to customers by 9/1/97 within a development cost of $700,000.

Purpose:
Strategy: To pursue a focus strategy emphasizing integrated circuit controllers for personal computer peripherals. Focus on high growth areas.

Strategic Alignment: High
- Project ties with firm's overall strategy for high growth, PC peripheral controllers.
- Intensity of rivalry among competitors is very high, but our engineers have experience with developing controllers for CD-ROM drives and prototype DVD-ROM drives.

Financial Impact: Estimated *ROI* through 9/1/99 = 30%

Scope:
Deliverables:
- First production units by 9/1/97.
- Design, production, and test documentation delivered to Document Control department by 6/1/97.
- Test hardware and software delivered to Manufacturing by 6/1/97.

Measurables: See attached sheet of Objective Specifications.

Exclusions:
- Project cost of $700,000 does not include marketing and selling costs.
- Engineering project which ends 9/1/97 does not include marketing and selling the product. However, the costs and benefits of marketing and selling the product were estimated to calculate the overall *ROI*.

Figure 22.3 shows the work breakdown structure for the DVD-ROM controller project.

The purpose of determining the work breakdown structure is to translate the project into action. Break the project into tasks that are:

- Manageable—able to assign specific responsibilities
- Measurable—able to monitor progress
- Integrateable into the whole project.

The WBS is a way to communicate clearly to all levels of the organization what needs to be done on the project.

Once the work breakdown structure is defined, it can be used to formulate other elements of the project plan such as defining the team members, the responsibility matrix, and determining the project schedule and costs.

1.3. Project Team Members

Before assigning team members to the project, it is important for the project manager to understand the company's organization and how it operates. This leads to understanding of the project manager's role and how it relates to the role of the functional managers.

There are many variations of organizational structure, but a typical organization is shown in Figure 22.4. In this typical organization, there are functional managers. They manage people in specific functions such as Engineering, Manufacturing, or Marketing. The functional managers have functional authority over people within their group.

Overlaying the functional groups are project teams. The project teams are led by project managers. In this example, there is a project manager for Project A and a project manager for Project B. The project teams are made up of members from various functional groups. The project managers have project authority over the project team members.

Since there is a potential conflict between the project managers and functional managers vying for the same resource, project managers need to understand organizational roles in order to work successfully with this system.

Figure 22.3. Work Breakdown Structure Example

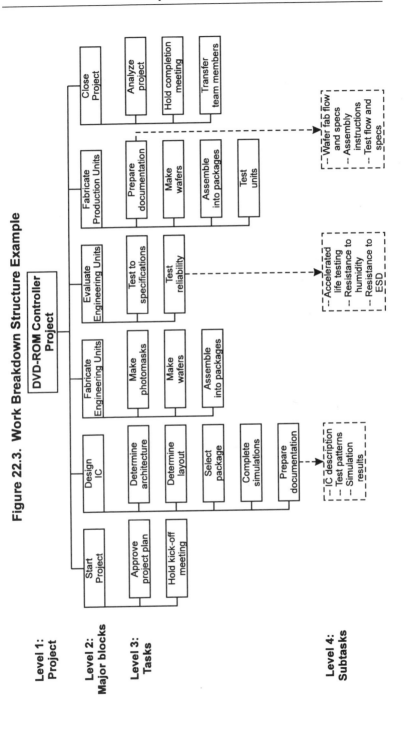

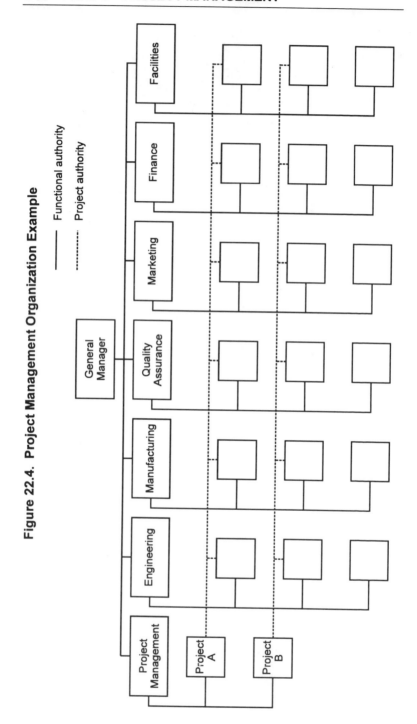

Figure 22.4. Project Management Organization Example

1.3.1. Project Manager's Role

The role of the project manager is to coordinate activities across functional groups in order to achieve the project objectives. The project manager guides the project through planning, implementation, monitoring, and closing of the project to ensure its success. He/she provides project leadership to the functional groups, communicating what must be done and by when. The project manager works with the functional manager to negotiate for resources.

1.3.2. Functional Manager's Role

Functional managers make commitments to do the work for the project. They determine who will do the work and how it will be done. Their group performs the work. Since the functional manager's group is involved with many projects, he or she works with the project managers in order to best allocate the functional resources according to priorities of the projects.

1.3.3. Executive's Role

In terms of the organizational example, the executives include the general manager and the manager of the project management group. One of the executive's roles is to participate actively during the evaluation phase of the projects and to select those projects that are desirable to sponsor through the planning phase. The executives help with project planning and setting objectives. Once the project plans have been completed, the executives decide which projects to implement.

During the implementation phase, the executives stay involved by setting priorities among the various projects. This provides a guideline for the functional managers to use when allocating resources. When conflicts still exist between project managers and functional managers, the executives help resolve the conflicts.

1.3.4. Defining the Team

In understanding how the organization works and the roles of various managers, it is clear that the project manager must work with the functional managers to determine who will be on the project team. Define what needs to be done by setting the project objectives and determining the work breakdown structure. Define the WBS with the functional managers. At this point, the functional managers will have a good understanding of the project and will be in a position to select who to put on the project team from their respective functional groups.

1.4. Responsibility Matrix

Up to this point, the project plan consists of the project objectives, the work breakdown structure, and a list of project team members. The next step is to blend the WBS and team member list together to form a responsibility matrix. Recall one of the purposes of the WBS is to break the project down into tasks whose responsibilities are readily assignable. The responsibility matrix is a graphical way to assign and communicate these responsibilities.

Figure 22.5 shows a responsibility matrix form, and Figure 22.6 shows this form filled out for the DVD-ROM controller project. The project team members are listed in the left hand column. The tasks defined in the WBS are listed along the bottom row. Thus a matrix is constructed consisting of the project team members and project tasks. Draw a solid dot at the intersection where a team member is responsible for a task. Draw a circle at the intersection where a team member contributes to a task.

By constructing a responsibility matrix, it is simple to visualize who is responsible and who contributes to complete each task. It is also easy to see with which tasks each team member is involved.

1.5. Project Schedule

The purpose of estimating the project schedule is to determine how to achieve the project objectives by balancing time, cost, and risk factors. Other reasons for estimating the project schedule are to communicate project activities in a graphical way, making sure resources are being used efficiently, and to make it easier to monitor the project.

There are two popular methods to estimate project schedules. One is the Program Evaluation and Review Technique (PERT), and the other is the Gantt (bar) chart.

Figure 22.5. Responsibility Matrix Form

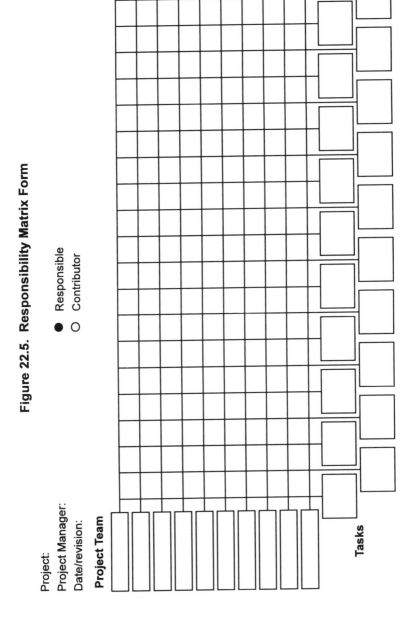

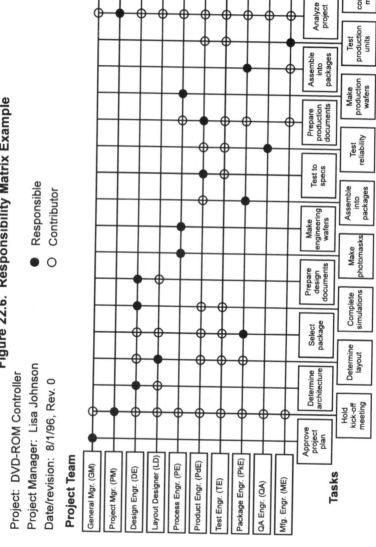

Figure 22.6. Responsibility Matrix Example

1.5.1. Program Evaluation and Review Technique (PERT)

PERT was originally developed in 1958-59 by the U.S. Navy with the help of the consulting firm Booz, Allen, and Hamilton. It was used with the Polaris Weapon System project.

An example of a PERT chart is shown in Figure 22.7. A feature of the PERT chart is the individual tasks are shown in boxes, and they can be visualized as a network. The network shows the relationships among the tasks. Within each task box, there is an estimate of the task duration. These time estimates are used to calculate the critical path for the project. The critical path is the sequence of tasks requiring the greatest expected time. The critical path is shown in the PERT chart by the bold task boxes connected by bold arrows. In the example shown in Figure 22.7, the critical path is going through tasks 1-2-3-6-7 because this takes 17 weeks. The alternate path of tasks 1-2-4-5-6-7 takes only 15 weeks.

Figure 22.7. PERT Chart

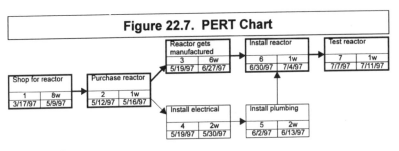

Assumptions:
1) Reactor delivery date = 6 weeks after purchase
2) Reactor can be installed in existing facility--
   only electrical and plumbing upgrades necessary

Some advantages of the PERT chart include the task relationships can be shown graphically in a network to help understand the overall project. Also changes in the project schedule can be evaluated easily by modifying the PERT chart.

A disadvantage using PERT charts is the chart can quickly become complex as the number of tasks increases. Also it is difficult to show the chart on a time scale. The task durations and calendar dates need to be read from the task boxes.

Table 22.2. Steps to Construct a PERT or Gantt Chart

1. List project tasks (take from WBS).

2. Make table placing tasks in order and identify task relationships.

3. Review table with functional managers for correctness.

4. Estimate time for each task.

5. Determine critical path.

6. Construct network diagram (for PERT chart) or bar chart (for Gantt chart) and convert time estimates into calendar dates.

7. Review PERT or Gantt chart with functional managers—iterative process.

8. List assumptions.

Table 22.3. Tasks for DVD-ROM Controller Project

Task	Description	Predecessor	Duration (weeks)
1	Approve project plan		0
2	Hold kick-off meeting	1	1
3	Determine architecture	2	8
4	Determine layout	3	4
5	Select package	4	2
6	Complete simulations	4	4
7	Prepare design documentation	5, 6	2
8	Make photomasks	7	1
9	Make engineering wafers	8	8
10	Assemble wafers into packages	9	2
11	Test to specifications	10	4
12	Test reliability	10	6
13	Prepare production documentation	11	2
14	Make production wafers	13	8
15	Assemble production wafers into packages	14	2
16	Test production units	12, 15	1
17	Analyze project	16	2
18	Hold project completion meeting	17	0
19	Transfer team members to other projects	16	2

The steps for constructing a PERT chart are shown in Table 22.2. These are also the same steps to construct a Gantt chart. First take the tasks identified in the WBS and list them. Make a table putting these tasks in order and define the task relationships. An example of such a table is shown in Table 22.3 for the DVD-ROM controller project. The predecessor column identifies which tasks need to be completed before a particular task can be started. Once you have completed this table, review it with the functional managers to see if any modifications need to be made. Then get time estimates. The best case for estimating the time for each task is to use historical data from similar tasks of other projects. If there is no historical data, get estimates from the functional managers. Next use project management software to determine the critical path, construct a network diagram, and convert time estimates to calendar dates. Review the PERT chart with the functional managers, and make necessary corrections. A final hint for using PERT charts is to include on your chart a list of assumptions used to make the schedule. In this way, people can quickly grasp the schedule estimates and assumptions without having to look elsewhere.

Once completed, the PERT chart shows the overall project schedule. It identifies the critical path and shows the relationships among tasks. As changes occur, the PERT chart can be modified to update the schedule.

Figure 22.8 shows the PERT for the DVD-ROM controller project. The overall project is estimated to take 49 weeks (estimated completion is 8/8/97). The critical paths are shown to be all tasks except for tasks 5 and 12. Any changes in schedule along the critical path will result in changes in the overall project schedule.

1.5.2. Gantt Chart

The Gantt chart is another method for estimating overall project schedules. It was developed by Henry Gantt in the early 1900s. The main feature of a Gantt chart is each task is represented by a bar which is plotted against time.

Figure 22.9 shows an example of a simple Gantt chart. This is the same set of tasks that was depicted in a PERT chart in Figure 22.7. In a Gantt chart, there is a table which lists each task, the estimated task duration, and the task predecessors. Then the timeline representing each task is plotted as a bar. As you can see in Figure 22.9, the overall project

Figure 22.8. PERT Chart for DVD-ROM Controller Project

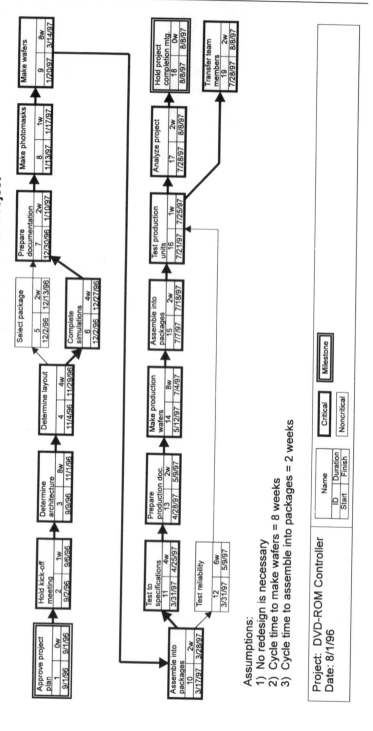

Assumptions:
1) No redesign is necessary
2) Cycle time to make wafers = 8 weeks
3) Cycle time to assemble into packages = 2 weeks

Project: DVD-ROM Controller
Date: 8/1/96

Figure 22.9. Gantt Chart

ID	Task Name	Duration	Start	Finish	Predecessors
1	Shop for reactor	8w	3/17/97	5/9/97	
2	Purchase reactor	1w	5/12/97	5/16/97	1
3	Reactor gets manufactured	6w	5/19/97	6/27/97	2
4	Install electrical	2w	5/19/97	5/30/97	2
5	Install plumbing	2w	6/2/97	6/13/97	4
6	Install reactor	1w	6/30/97	7/4/97	3, 5
7	Test reactor	1w	7/7/97	7/11/97	6

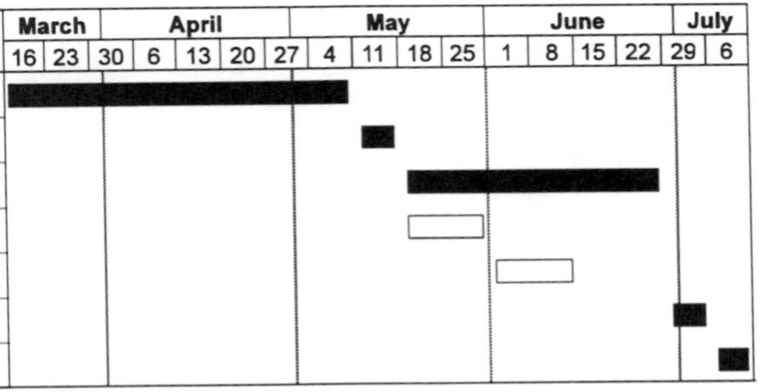

Assumptions:
1) Reactor delivery date = 6 weeks after purchase
2) Reactor can be installed in existing facility--only electrical and plumbing upgrades necessary

Project: New Reactor   Critical ▮   Noncritical ▯
Date: 3/1/97

schedule is still 17 weeks, and the critical path to complete the project is still through tasks 1-2-3-6-7.

The Gantt chart shares some of the advantages with the PERT chart. The Gantt chart is a way to depict graphically the overall project schedule, and changes to the schedule can be evaluated easily by modifying the chart.

The main advantage of the Gantt chart over the PERT chart is that tasks can be plotted on a time scale. Also the Gantt chart can handle a greater number of tasks without the chart becoming complex.

The main disadvantage of the Gantt chart is that task relationships may not be as clear visually as they are in a PERT chart. In a Gantt chart, the task predecessors are listed in a table, but they are not shown as a network.

The steps for constructing a Gantt chart are virtually the same as constructing a PERT chart. The Gantt chart steps are shown in Table 22.2. Use the same table of tasks which lists the relationships and time estimates. The only difference is to command your project management software to plot a Gantt bar chart instead of a PERT chart. Again it is important to list your scheduling assumptions right on the chart.

For the DVD-ROM controller project, use Table 22.3 as the table of tasks. The corresponding Gantt chart is shown in Figure 22.10. The information presented in terms of overall project schedule is the same as using a PERT chart. Which chart you use for your project depends on your preferences. Use a Gantt chart if showing the schedule on a time scale is more important. Use a PERT chart if showing the task relationships is more important.

1.6. Project Costs

To estimate the overall project costs, begin with the list of tasks from the WBS. Estimate the cost of each task. There are two types of costs to consider: labor and non-labor costs. To determine labor costs, estimate the labor hours necessary to complete the task and multiply by the appropriate labor rate. Reference the responsibility matrix and the task durations when estimating labor costs to remind yourself who is working on each task. Non-labor costs can consist of materials, equipment, and services necessary to complete the tasks.

Figure 22.10. Gantt Chart for DVD-ROM Controller Project

ID	Task Name	Duration	Start	Finish	Predecessors
1	Approve project plan	0w	9/1/96	9/1/96	
2	Hold kick-off meeting	1w	9/2/96	9/6/96	1
3	Determine architecture	8w	9/9/96	11/1/96	2
4	Determine layout	4w	11/4/96	11/29/96	3
5	Select package	2w	12/2/96	12/13/96	4
6	Complete simulations	4w	12/2/96	12/27/96	4
7	Prepare documentation	2w	12/30/96	1/10/97	5, 6
8	Make photomasks	1w	1/13/97	1/17/97	7
9	Make wafers	8w	1/20/97	3/14/97	8
10	Assemble into packages	2w	3/17/97	3/28/97	9
11	Test to specifications	4w	3/31/97	4/25/97	10
12	Test reliability	6w	3/31/97	5/9/97	10
13	Prepare production doc.	2w	4/28/97	5/9/97	11
14	Make production wafers	8w	5/12/97	7/4/97	13
15	Assemble into packages	2w	7/7/97	7/18/97	14
16	Test production units	1w	7/21/97	7/25/97	12, 15
17	Analyze project	2w	7/28/97	8/8/97	16
18	Hold project completion mtg.	0w	8/8/97	8/8/97	17
19	Transfer team members	2w	7/28/97	8/8/97	16

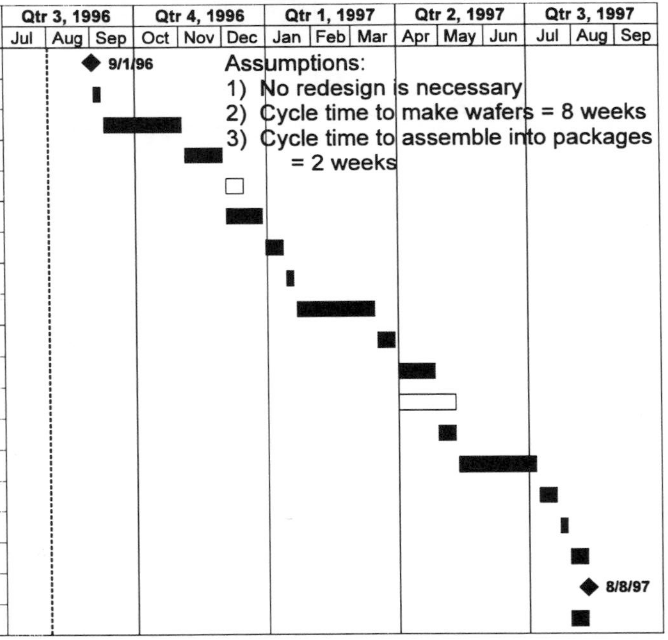

Assumptions:
1) No redesign is necessary
2) Cycle time to make wafers = 8 weeks
3) Cycle time to assemble into packages = 2 weeks

Project: DVD-ROM Controller
Date: 8/1/96

Critical ■   Noncritical ▢   Milestone ◆

As was the case with estimating schedules, the most accurate way to estimate costs is to use historical data. Look for instances where similar tasks were performed for other projects. Work with the functional managers to estimate costs.

Table 22.4 shows an example of determining the project costs for the DVD-ROM controller project. The labor and non-labor costs were estimated for each task. The total project cost is estimated to be $619,600.

Table 22.4. Costs for DVD-ROM Controller Project

Task	Description	Hours	Labor Rate ($/hr.)	Labor cost	Non-labor cost	Total Task Cost
1	Approve project plan			$0	$0	$0
2	Hold kick-off meeting	40	$100	$4,000	$0	$4,000
3	Determine architecture	640	$100	$64,000	$0	$64,000
4	Determine layout	480	$100	$48,000	$0	$48,000
5	Select package	80	$100	$8,000	$0	$8,000
6	Complete simulations	160	$100	$16,000	$0	$16,000
7	Prepare design documentation	80	$100	$8,000	$0	$8,000
8	Make photomasks	40	$100	$4,000	$40,000	$44,000
9	Make wafers	80	$100	$8,000	$70,000	$78,000
10	Assemble wafers into packages	20	$100	$2,000	$20,000	$22,000
11	Test to specifications	320	$100	$32,000	$140,000	$172,000
12	Test reliability	240	$100	$24,000	$40,000	$64,000
13	Prepare production documentation	160	$100	$16,000	$0	$16,000
14	Make production wafers	80	$100	$8,000	$35,000	$43,000
15	Assemble production wafers into packages	10	$100	$1,000	$10,000	$11,000
16	Test production units	16	$100	$1,600	$0	$1,600
17	Analyze project	160	$100	$16,000	$0	$16,000
18	Hold project completion meeting	40	$100	$4,000	$0	$4,000
19	Transfer team members to other projects			$0	$0	$0
	TOTALS	**2646**		**$264,600**	**$355,000**	**$619,600**

Use project management software to graph the project costs over time. A useful way of doing this is to plot the cost by period and cumulative costs alongside the Gantt chart. This is shown for the DVD-ROM controller project in Figure 22.11. Also included in this figure is a listing of the task predecessor relationships and who is responsible for each task. This is a very effective way to highlight elements of the project plan in one picture: the work breakdown structure, responsibilities, schedule, and costs.

1.7. Project Closure

The final element of the project plan is a plan to close the project. A complete project plan outlines not only how the project objectives will be achieved, but it also outlines what will happen once the objectives are achieved.

Figure 22.11. Cost and Schedule for DVD-ROM Controller Project

Project Plan
  Start: 9/1/96
  Complete: 8/8/97
  Cost: $619,600

Assumptions
  1) No redesign is necessary
  2) Cycle time to make wafers = 8 weeks
  3) Cycle time to assemble wafers into packages = 2 weeks

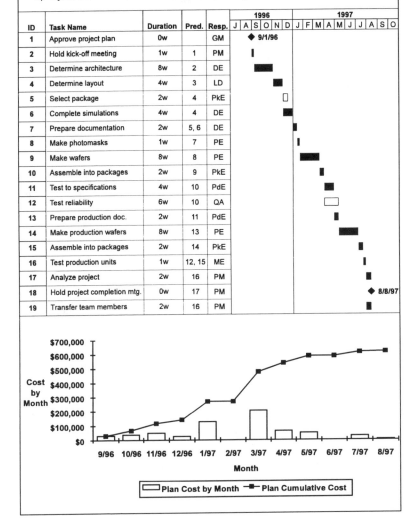

ID	Task Name	Duration	Pred.	Resp.	1996 / 1997 schedule
1	Approve project plan	0w		GM	◆ 9/1/96
2	Hold kick-off meeting	1w	1	PM	
3	Determine architecture	8w	2	DE	
4	Determine layout	4w	3	LD	
5	Select package	2w	4	PkE	
6	Complete simulations	4w	4	DE	
7	Prepare documentation	2w	5, 6	DE	
8	Make photomasks	1w	7	PE	
9	Make wafers	8w	8	PE	
10	Assemble into packages	2w	9	PkE	
11	Test to specifications	4w	10	PdE	
12	Test reliability	6w	10	QA	
13	Prepare production doc.	2w	11	PdE	
14	Make production wafers	8w	13	PE	
15	Assemble into packages	2w	14	PkE	
16	Test production units	1w	12, 15	ME	
17	Analyze project	2w	16	PM	
18	Hold project completion mtg.	0w	17	PM	◆ 8/8/97
19	Transfer team members	2w	16	PM	

Cost by Month chart — axis: $0 to $700,000; months 9/96 through 8/97.

Legend: □ Plan Cost by Month   ■ Plan Cumulative Cost

Items to consider in the closure plan include how to give recognition to the team members in terms of awards or career advancement. Plan out how the team members will be transferred to other projects. Determine how and to whom project information will be transferred. Write a final project report to summarize accomplishments. Analyze what went well and what could have been done better in an effort to help future projects. Hold a closure meeting to recognize the contributions of everyone involved in the project.

2. PROJECT PLAN GUIDELINES

Chapter 22 up to this point has described the key components of the project plan—the desired outputs of the planning cycle. The rest of the chapter is devoted to project plan guidelines which advises you how to go about completing the project plan.

Effective communication is the key. Use all the communication skills you learned in Part I. Take the time to think about the project. When communicating with others, practice the principles of effective interpersonal communication, emphasize listening to understand the position of others, and be prepared and focused when presenting your ideas. Spend the time to maintain constructive relationships with your boss and your co-workers.

In addition to these communication skills, there are other specific skills to use as you lead the project planning phase. These are gaining support from the functional managers and gaining support from the project team.

2.1. Gaining Support from Functional Managers

Recall from the discussion about roles that the project manager has overall responsibility for the project. You coordinate the activities of project team members from various functional groups in order to attain the project objectives. The functional managers on the other hand, make commitments to do the project work. They determine who will do the work and how it will be done. Thus the project manager has a lot of responsibility but in some cases not much authority regarding resource assignments and commitments. The project manager must win the support from various functional mangers in order to get the necessary resources for the project.

**Table 22.5. Keys to Gaining Support
from Functional Managers**

1. Establish constructive relationships.
2. Show how involvement in the project will benefit both of you.
3. Address functional manager's concerns.
4. Request specific support.
5. Agree on a specific follow-up plan.

Winning the support from functional managers is critical during the project planning phase. Close work and agreement with the functional managers is necessary to determine the work breakdown structure, define the team members, set the responsibility matrix, and estimate the project schedule and costs.

The keys to winning support from the functional managers are shown in Table 22.5. As a project manager, you must continually sell yourself, your project, and your ideas to others. Thus you need successful sales skills to be an effective project manager. Keep in mind successful selling involves understanding the needs and wants of the other person. Then act to meet the needs of that person first and your needs second.

The first key is to establish constructive relationships with the functional managers. Ideally you would have established these constructive relationships before the start of the project. If you have not done this yet, take the time to talk individually with the functional managers, explore common points of interest, and identify ways to help each other.

The second key is to discuss the project in terms of how it will benefit both of you as well as the organization. Recognize the involvement and contributions of the functional managers as often as you can throughout the project. Arrange to have the customers of the project meet with the functional managers, so the functional managers can experience directly what type of impact they will have.

Next address the concerns of the functional manager. After reviewing the benefits of the project, check for understanding and reactions from the functional manager. The concerns need to be resolved before there will be much support. Use your listening skills to make sure you understand the concerns. Once the concerns are understood, adopt an approach to solve any problems together. Invent solutions that will be beneficial to both of you. Addressing the concerns of the functional manager will take a lot of time, but it is time well spent. Keep emphasizing the benefits of the project to the functional manager.

The next key is to request specific support from the functional manager and their group. This support should be what is outlined in the work breakdown structure and responsibility matrix. And finally, agree on a specific plan for support from the functional manager. Arrange for a follow-up meeting with the functional manager to discuss progress towards supporting the project. Express appreciation for any commitments you receive.

2.2. Gaining Support from the Project Team

As you go through the project planning phase, the project team members are determined. It is important for you as the project manager to provide clear and effective leadership. In the short-term, this helps to develop a sound project plan to which the project team is committed. In the long-term, effective leadership aids in achieving the project objectives. Gaining support from team members is just as vital as gaining support from functional managers. Involve the team when determining the work breakdown structure, setting the responsibility matrix, and estimating the project schedule and costs. Without the team's support, you will not have commitment. Engineers with the ability to lead project teams are valued at a premium by companies.

The keys to winning support from the team members are shown in Table 22.6. The first step is to communicate a clear vision of what needs to be done and show why it is important for the organization. This helps to solidify the project team's purpose and importance. If the team members can understand the team's purpose, it will be easier for them to adopt the team's purpose as their own. Use the Project Objectives form as a basis to begin the discussion. As you present this, ask the team members for their reactions, so you can gauge their level of commitment. Request the help of executives who selected the project to aid in communicating the project's purpose.

<table>
<tr><td colspan="2" align="center">**Table 22.6. Keys to Gaining Support
from Project Team**</td></tr>
<tr><td>1.</td><td>Communicate the project objectives and how they impact the organization.</td></tr>
<tr><td>2.</td><td>Clarify roles and responsibilities for each team member.</td></tr>
<tr><td>3.</td><td>Define guidelines and resources to get the job done.</td></tr>
<tr><td>4.</td><td>Agree on a specific follow-up plan.</td></tr>
</table>

Next clarify the roles and responsibilities for each team member. Unclear roles and responsibilities are a major cause of people problems and lead to destruction of the team's effectiveness. Involve the team as you develop the work breakdown structure and responsibility matrix. Try to continually describe the roles and responsibilities in terms of how they contribute to the team's purpose. Once established, review periodically the member's roles and responsibilities in order to keep them clear.

Define the guidelines and resources to get the job done. This step shows how to set the project plan into action. Set guidelines to help determine such things as:

- how decisions will be made

- how priorities are determined

- how to communicate within the team

- how to ask for help

Encourage the team's involvement to develop the guidelines, but be firm on guidelines that are not negotiable. An example of a non-negotiable guideline is work practices relating to safety. Be prepared with reasons why certain items are not negotiable. Work toward obtaining agreement among team members on guidelines that are negotiable. While developing the guidelines, encourage the team members to find ways to help each other. This not only enhances teamwork, it also helps you accomplish more with the same amount of resources.

Finally agree on a specific follow-up plan. The first item of this plan should be to summarize what was discussed and agreed to concerning the first three items: the project objectives, roles, and guidelines. Write these down and distribute them to each team member. Set a date for continued discussions about the project plan.

Working with the project team to win their support takes time. In the short-term, this delays completing the planning phase and getting to the implementation phase. But in the long-term, time and effort here lays the groundwork for getting commitment and getting the team motivated to work on the project. And this will result in faster accomplishment of the project objectives.

3. PROJECT PLAN SUMMARY

The project plan chapter has described to you key elements of a successful project plan and guidelines of how to work with functional managers and the project team in order to get the project plan done.

Keep in mind project planning is an iterative process. Upon completing the first pass of the plan, it is usually necessary to modify and refine the plan to come up with a plan that is designed to achieve the project objectives. More specifically, it is usually necessary to figure out ways to complete the project in a shorter amount of time and with less resources.

One of the purposes of the project plan is to identify problem areas before they occur. Once identified, you need to take corrective action to solve the problems. In some cases, you may consider developing backup plans in case certain problems surface during the course of the project.

Once the project plan is completed, submit it to the executives. It is the executives' responsibility to decide which projects to carry out. They need to commit the necessary money and resources for the project. If your project is approved, the next phase of the project is the implementation phase.

Chapter 23

PROJECT IMPLEMENTATION

So far we have covered the first two phases of a project: evaluation and planning. In the evaluation phase, executives select which projects to carry through to the planning phase. After the planning phase, executives decide whether or not to authorize the necessary resources. Projects that pass these first two phases then go on to the implementation phase.

As the project manager, understanding the first two phases makes it easier to lead the project implementation. By understanding the findings of the evaluation phase, you know the objectives of the project and how it fits in with the organization's larger picture. Communicating this vision during the planning phase helps you win support from others. Developing the key elements of a project plan points out how all the tasks fit together in order to achieve the project objectives. The plan also highlights where there may be problem areas and gives you a chance to proactively take corrective action.

During the implementation phase, you set the project plan into action. Even with the best laid project plans, problems and conflicts will surface. An important part of your success as a project manager will be determined by how well you can resolve conflicts. This chapter will give you tools how to resolve conflicts and solve problems. Another important part of the implementation phase is holding effective meetings with your project team, the executives, and your customers. How to run meetings will be discussed. Also traits of a successful project manager will be discussed in order to help guide you throughout the project management experience.

1. TRAITS OF A SUCCESSFUL PROJECT MANAGER

As your project gets to the implementation phase, it is important for you to stay on track as a project manager. Table 23.1 lists traits to emulate.

Portray a clear direction to all those involved in the project. Having a clear direction is one of the benefits of thorough planning. In addition, you need to maintain this clear sense of direction throughout the life of the project to keep the team focused on the objectives.

Focus on team building and continue practicing strong interpersonal communication skills. Team building during the initial formation of the team was discussed in the project planning chapter. Continue to involve the team with decisions when possible and maintain open communications with the team. Strong interpersonal skills continue to be important since the process of planning and getting commitments never ends. Covey says an effective way to work with your team on a continual basis to handle any task is to:

- communicate the desired results
- define who is accountable for the task
- identify the consequences (positive consequences if successful, negative consequences if not successful)
- outline the guidelines for completing the task[57]

In this way you empower your project team to perform the tasks. This is an effective way to build the project team. Your main role becomes steering the project team in the right direction and resolving conflicts when they arise. Another vehicle for building teams and communicating the status of the project is the project review meeting. Running meetings will be described in the next section.

Be action oriented. This is especially crucial in the implementation phase because the plan is set into action. Utilize resources effectively to get results.

---

57. Stephen R. Covey, *Principle-Centered Leadership,* New York: Simon & Schuster, 1991.

Table 23.1. Traits of a Successful Project Manager
1. Clear direction with a sound plan
2. Team builder
3. Strong interpersonal communication skills
4. Action oriented
5. Resolves conflicts—solves problems
6. Dedicated to quality and the customer
7. Balances technical, economic, and human factors

Take an active role to resolve conflicts and solve problems. Resolving conflicts will be discussed in the last section of this chapter.

Remain dedicated to quality and the needs of the customer. Utilize statistics whenever possible to quantify aspects of quality. Build in quality from the outset of the project in order to save time in the long run. Maintain close communications with the customer and seek feedback from them throughout the life of the project.

Balance technical, financial, and human factors. Technical aspects will always be crucial when managing an engineering project. However technical factors need to be balanced with financial and human factors. A primary aim of this book is to give you tools and knowledge to maintain this balance. Stay consistent and stable in reaching for the project objectives, but be flexible and adaptable enough to work through the problems that arise.

2. HOW TO RUN A MEETING

Project review meetings are often the most efficient way of coordinating activities of the team and to communicate the status of the project. The pace of the work proceeds so quickly that meetings are often the best way to find out what is happening. Meeting regularly helps define the team and build teamwork. Discussing project topics in meetings increases commitment to the decisions made. Also, we are a social species—meetings provide a way to communicate face-to-face.

Because of the social aspects, meetings will never be completely replaced by new technology.

However, with all the potential benefits, many meetings deteriorate into a waste of time. The problems with many meetings are that they are unnecessary (held for traditional reasons) and poorly organized. The purpose of this section is to provide guidelines to run successful meetings.

2.1. Types of Meetings

While managing a project, there are three types of meetings:

- project team review meetings
- executive review meetings
- customer review meetings

The project team review meetings are the most informal of the three. They are called whenever viewed to be beneficial—this may be every week or two weeks. Project team review meetings help the members do their work more intelligently and increase the efficiency of communications. Regular project team review meetings help maintain and reaffirm the team's collective aim and how the members contribute towards it. At these meetings, have the responsible person for the appropriate tasks (defined by the responsibility matrix) discuss progress made toward those tasks. Note the problems and discuss them if appropriate.

Executive review meetings are meetings with those people who originally decided to sponsor and support the project. Attendees may include the president, vice president, general manager, and functional managers of the organization. These meetings are more formal than the project team review meetings. The executive review meetings are usually held monthly. The primary objective of the meeting is to communicate the status of the project and where there may be issues limiting progress. It is your responsibility to keep the executives well-informed about the project. The more meaningful you make these status reports, the less likely the executives will meddle with the project. During the implementation phase, the executives' primary role is to set priorities and to help resolve conflicts when they can't be resolved by the team. During this phase, executives should step back and let you

Table 23.2. Guidelines to Run a Successful Meeting
1. Make sure all meeting attendees know the purpose and outcomes expected of the meeting.
2. Prepare participants by stating before the meeting what you need from them.
3. Be punctual.
4. Encourage two-way communication.
5. Keep meeting focused on its purpose and desired outcomes.
6. Summarize key points and write meeting minutes.

run the project. But always remember your responsibility to keep them well-informed.

The third type of meeting is the customer review meeting. This is the most formal of the meetings and often the most critical. Doing an effective job of communicating project progress and issues at these meetings will go a long way towards building trust with the customer. High trust eases all other communications with the customer.

2.2. Guidelines

As discussed previously, you gain a number of benefits by running effective project review meetings. Even though there are three types of meetings, the guidelines are similar for all three. These guidelines are listed in Table 23.2. Running an effective meeting takes preparation. Taking the time to prepare is a common theme throughout this book. Do not leave it up to chance to succeed as an engineer. Make time to prepare and plan.

The most important question to ask about a meeting is, "What is the purpose and the desired outcome?"[58] Once these are determined, write up an agenda. Every item on the agenda should help achieve the meeting objective. Be wary of trying to accomplish too much with the

---

58. Antony Jay, "How to Run a Meeting," *Harvard Business Review,* March-April 1976, 43.

meeting. Keep the meeting objective simple and the agenda simple. Try to make the agenda descriptive enough to eliminate ambiguity. You may even want to include after each agenda item one of the following descriptions:

- for information
- for discussion
- for decision

This alerts the meeting attendees how you plan to cover each item during the meeting. Distribute the agenda two or three days in advance of the meeting. If you distribute it any earlier, the participants may lose it or forget about the meeting. If you distribute it any later, the participants may not have enough time to prepare for the meeting.

The second guideline has to do with preparing participants for the meeting. Sit down and talk face-to-face with key participants before the meeting to make sure they understand the issues that will be discussed. If you need them to present information at the meeting, explain exactly what you need and why it is important. Give them adequate time to prepare any presentations.

Be punctual. Put the start and finish times on the agenda. Then reward those who show up on time by abiding by those times. Start the meeting by clearly stating the purpose and the desired outcome of the meeting. Show the agenda. Few meetings achieve much after two hours. For meetings that tend to go long, try starting them an hour before lunch or near the end of the work day. Then the attendees will be motivated to end the meeting in a timely manner so it will not conflict with lunch or going home.

The fourth guideline is encourage two-way communication. Information needs to be exchanged at these project review meetings. The best way to accomplish this is through active participation by the meeting attendees. Encourage two-way communication by expressing appreciation for constructive participation. Seek contributions from quiet attendees by calling on them. Keep an open mind to the points brought up by the participants. The most meaningful meetings are when the attendees get involved. One of the clearest danger signals is hearing yourself talk a lot during the meeting. Be wary of coming on too strongly during the discussion with your own ideas and lecturing.

A primary role you have while running the meeting is to keep it focused on its purpose. Even though you don't want to dominate the discussion, you are the person to keep the meeting on track. Do this by restating the purpose of the meeting if it gets off track. Exert control over any individual who is in danger of diverting progress toward the meeting objectives. Serve the group by clarifying points when necessary, and keep the discussion moving forward. Work through the agenda in a logical fashion. Try to discuss each agenda item thoroughly before moving on. This helps you avoid skipping backwards and forwards through the agenda which muddles the meeting's sense of direction.

Conclude the meeting by summarizing the key points and how it relates to the original purpose and desired outcome. There are two reasons for doing this. One is to give the meeting attendees a sense of accomplishment for their participation. The second reason is to check for understanding. The key points you summarize at the close of the meeting may not be the same as what the participants understood. Summarizing the key points gives them an opportunity to clarify any misunderstandings that may exist. To help summarize the key points, write down key points and action items assigned during the course of the meeting. Do this on a board, flip chart, or overhead transparency so that it is visible to all the meeting attendees.

Finally write up meeting minutes, and distribute them to the participants. Forty percent of what is said is forgotten after half an hour and ninety percent is forgotten after a week. It is important to write down what transpired at the meeting before it is forgotten. To help write the minutes, consider assigning someone at the meeting to take notes for you. Many times it is difficult to run the meeting and take notes at the same time. Include in the minutes the meeting's purpose, the key points, and action items assigned.

3. RESOLVING CONFLICTS

Conflicts are a way of life for project managers. Thorough planning can help anticipate many potential sources of conflict before they occur, but conflicts can never be eliminated. One of the measures of how successful you are as a project manager is your ability to resolve conflicts.

Conflicts are not all bad. They are an essential part of any organization. Conflicts can spark creativity, provide motivation to be innovative, and encourage personal improvement. However, if a conflict is adversely impacting your project team's performance, you need to take an active role to solve the problem.

Over the life of the project, the most intense conflicts come from disagreements over:

- schedules

- project priorities

- manpower resources

- technical opinions

There may also be conflicts over costs and administrative procedures such as reporting structure and communications. A main reason for these conflicts is you have limited control over the project resources. You have a lot of responsibility, but in many cases you have limited authority. Thus you need to use your negotiating skills to gain resources from the various functional departments.

3.1. Conflict Resolution Modes

There are many ways to deal with a conflict. Figure 23.1 shows a model of various conflict resolution modes from Ruble and Thomas. This shows five modes that can be classified by two dimensions: cooperativeness and assertiveness. Cooperativeness is the degree to which you try to satisfy the concerns of the other party. Assertiveness is the degree to which you try to satisfy your own concerns. You may have to use all five of these approaches at various times depending on the situation.[59] After describing the five resolution modes, we will concentrate on the collaborative approach.

Forcing. With this mode you force your own solution to resolve the conflict. This is the uncooperative and assertive approach. This approach shows a lack of tolerance and self-confidence. Repeated use of this approach can breed resentment among the people you work with. Forcing is most likely to be used when the time pressure is great, you

---

59. David A. Whetten, and Kim S. Cameron, *Developing Management Skills,* Glenview, Illinois: Scott, Foresman and Company, 1984.

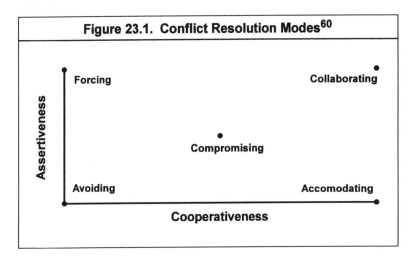

Figure 23.1. Conflict Resolution Modes[60]

are the supervisor of the other party, and maintaining a supportive relationship is not important.

Avoiding. Avoiding is the uncooperative, unassertive approach. By avoiding the conflict, this is a sign you are ill-equipped to resolve the conflict. Repeated use of this approach will create much frustration for your project team.

Accommodating. This is the cooperative, unassertive approach. Accommodating emphasizes preserving a friendly relationship at all costs. This results in people taking advantage of you, lowering your self-esteem. Accommodating is likely to be used when the conflict involves a formal subordinate/supervisor relationship and when it is important to maintain a supportive relationship.

Compromising. This is the attempt to partially satisfy both parties. In the conflict resolution model, compromising is halfway on the cooperativeness scale and halfway on the assertiveness scale. Handling a problem with this mode shows you are more interested in resolving the dispute rather than solving the problem. The end result is this encourages game playing on the part of the participants to get what they want.

60.   T. Ruble, and K. Thomas, "Support for a Two-Dimensional Model of Conflict Behavior," *Organizational Behavior and Human Performance,* 1976, 16, 145.

<u>Collaborating.</u> Collaborating is the conflict resolution mode high in cooperativeness and high in assertiveness. This approach emphasizes problem solving rather than finding blame. It focuses on the problem rather than personalities. Collaborating fosters an environment of trust while being cooperative and assertive in an effort to solve the problem.

Although collaborating is the superior conflict resolution mode, it is not appropriate for all situations. Collaborating requires time. If the time pressure is urgent, collaborating may not be possible. Collaborating is most useful when the time pressure is not urgent, the issue is critical or complex, and it is important to maintain a supportive relationship. The next section is devoted to give guidelines how to use this approach.

3.2. Guidelines to Use Collaboration

Collaborating is the most difficult conflict resolution approach. Not only do you need to be skillful and mature to go through the process, you are also taking risks with the amount of control you have over the situation. With a more authoritative approach such as forcing, it is safer for you in some ways. In the short-term, forcing maintains your control over the project, and it is an efficient way to resolve a conflict. With collaborating you involve people in resolving the conflict. As you involve people, you lose some of your direct control.

The payback of the collaborating approach is a more effective way to resolve a conflict. By involving people in the problem, they are more committed to the decision made. People are more interested in their own ideas rather than someone else's. Also you maintain constructive relationships by showing confidence in others to help resolve the conflict.

As you use the collaborative approach, keep in mind the principles for good interpersonal communication. Seek to understand the situation, and keep an open mind to debatable points. Meet face-to-face with those involved with the conflict if possible. Timing of this meeting is critical. Call this meeting as soon as you can in order to resolve the conflict before it becomes a larger issue.

Table 23.3 shows the steps involved with using the collaborative approach. The first step is to identify the problem. This is the most critical and difficult step. Promptly call a meeting with those involved. In this meeting, set up a joint problem solving approach by encouraging

Table 23.3. Steps to Use the Collaborating Approach to Resolve Conflicts

1. Identify the problem.
  • Describe to the people involved how the conflict is impacting performance.
  • Seek to understand all sides of the issue.
  • Agree on what is the problem.
2. Generate possible solutions from each side of the conflict.
3. Get commitment as to what will be done to resolve the conflict.
4. Summarize and set a date to check on progress.

two-way communication. Start the meeting by describing how the conflict is impacting the team's performance. This lets the people know why you are getting involved with the problem. Next use your listening skills to understand all sides of the issue. Ask the people involved to describe their viewpoint in an objective way. Focus on the situation, not on the personalities involved. A key to using the collaborative approach is to focus the parties involved to solve the problem and not become territorial or defensive. Avoid becoming too evaluative at this point. After hearing all sides, come to an agreement as to what is the problem. If you have difficulty reaching consensus, reemphasize the impact the conflict is having on performance. This reminds those involved the importance of coming to an agreement.

After identifying the problem, have everyone generate possible solutions. This is the strength of the collaborative approach. People are more likely to implement a solution and be committed to it if the ideas are their own. Once the possible solutions are on the table, guide the group to decide which solution to implement.

Get commitment from each person as to what will be done to solve the problem. You may have to review the conflict's impact on performance and what is the problem in order to increase the level of commitment.

Close the meeting by summarizing the problem and everyone's commitment towards resolving it. Record these items for later reference. Set a date to check on progress. This shows you will be holding people accountable for the commitments they made to resolve the conflict. Finally express your appreciation to the people at the meeting for working through the conflict to come up with a solution.

Chapter 24

MONITORING PROJECTS

During the implementation phase of the project, it is not sufficient to clearly communicate the project objectives. You also need to communicate the progress being made towards those objectives. This is the purpose of project monitoring.

One of your duties as a project manager is to provide frequent and meaningful status reports. Effectively communicating project status and resolving problems reduces the amount of micro-managing by executives. As the executives become less involved in the day-to-day details of the project, they can concentrate on their role to set overall long-term direction for the organization. Executives are also needed to set overall priorities among various projects and to be available to help resolve those conflicts that cannot be resolved within the project team.

Keep your project status reports short and concise. This shortens the feedback loop for any action that needs to be taken based on the report. Include only pertinent information, and try to maintain a consistent report format throughout the project.

Figure 24.1 shows an example of how to portray graphically the project status. This uses the project plan shown in Figure 22.11 and compares the actual task schedule and costs to the plan. At one glance you can see the status of the project. This is an effective way to present the project status at your project review meetings.

To supplement this graphical representation, include a one page report summarizing the schedule, costs, problems, and actions taken to correct the problems. An example of such a form is shown in Figure 24.2. This form is filled out for the DVD-ROM controller project in Figure 24.3.

Figure 24.1. Status of DVD-ROM Controller Project

Summary as of 12/1/96

	Plan	Actual/Projected
Start	9/1/96	9/1/96
Complete	8/8/97	8/22/97
Cost	$619,600	$635,600

Assumptions
1) No redesign is necessary **(Assumption invalid)**
2) Cycle time to make wafers = 8 weeks
3) Cycle time to assemble wafers into packages = 2 weeks

ID	Task Name	Duration	Pred.	Resp.	1996 / 1997
1	Approve project plan	0w		GM	◆ 9/1/96
2	Hold kick-off meeting	1w	1	PM	100%
3	Determine architecture	8w	2	DE	100%
4	Determine layout	4w	3	LD	50%
5	Select package	2w	4	PkE	0%
6	Complete simulations	4w	4	DE	0%
7	Prepare documentation	2w	5, 6	DE	0%
8	Make photomasks	1w	7	PE	0%
9	Make wafers	8w	8	PE	0%
10	Assemble into packages	2w	9	PkE	0%
11	Test to specifications	4w	10	PdE	0%
12	Test reliability	6w	10	QA	0%
13	Prepare production doc.	2w	11	PdE	0%
14	Make production wafers	8w	13	PE	0%
15	Assemble into packages	2w	14	PkE	0%
16	Test production units	1w	12, 15	ME	0%
17	Analyze project	2w	16	PM	0%
18	Hold project completion mtg.	0w	17	PM	◇◆ 8/22/97
19	Transfer team members	2w	16	PM	0%

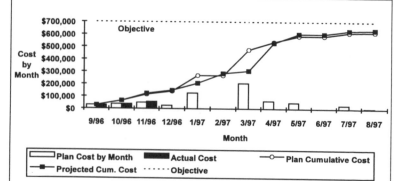

Legend:
☐ Plan Cost by Month ■ Actual Cost —○— Plan Cumulative Cost
—■— Projected Cum. Cost ····· Objective

Figure 24.2. Project Report Form

Project:
Project Manager:
Project Objective:
Date:

Schedule Summary:

	Plan	Projected
Project Completion Date		

Tasks Completed Last Month	Original Plan Date	Last Month Plan Date	Actual Date

Tasks for Next Month	Original Plan Date	Current Plan Date

Cost Summary:

	Plan	Projected
Cost to Complete Project		

	Plan	Actual
Costs to Date		

Problems	Potential Impact	Corrective Action

Comments:

Figure 24.3. Project Report Form for DVD-ROM Controller

Project: DVD-ROM Controller
Project Manager: Lisa Johnson
Project Objective: 1st production units by 9/1/97 within a development
                      cost of $700,000
Date: 12/1/96

Schedule Summary:

	Plan	Projected
Project Completion Date	8/8/97	8/22/97

Tasks Completed Last Month	Original Plan Date	Last Month Plan Date	Actual Date
Determine architecture	11/1/96	11/15/96	11/15/96

Tasks for Next Month	Original Plan Date	Current Plan Date
Determine layout	11/29/96	12/13/96
Select package	12/13/96	12/27/96

Cost Summary:

	Plan	Projected
Cost to Complete Project	$619,600	$635,600

	Plan	Actual
Costs to Date	$116,000	$124,000

Problems	Potential Impact	Corrective Action
Customer changed specifications during design phase	• 2 wk. delay to determine architecture, leading to 2 wk. delay in overall project. • $16,000 increase in overall costs.	• Customer agreed to delay and will pay for added cost. • Obtained guarantee from customer there will be no more specification changes for this project.

Comments:
- Customer change violated assumption that no redesigning would be necessary.
- Even with added work due to customer specification change, the project is still on target to meet the original project objectives: 1st production units by 9/1/97 and within a cost of $700,000.

A key part of monitoring the project is how you take corrective action to solve the problems that arise. The last chapter covered how to resolve conflicts. Use these techniques to take corrective action. Strive to have a corrective action for each of the problems listed in the report.

Recall the depiction of project phases shown in Figure 20.1. The five phases are evaluation, planning, implementation, monitoring, and closing. The planning/implementation/monitoring phases are iterative. Even with the best laid out plans, there will always be changes made while implementing the plan. By doing a sound job of monitoring the project, you will quickly recognize problems when they occur. Take corrective action promptly. Make changes to the plan if necessary.

Chapter 25

CLOSING PROJECTS

Successful projects have a well-defined closing phase that is planned long before the project ends. Throughout the project, say what you are going to do, then do it. The project close is no exception. There are several aspects to consider to complete the project.

Analyze how the project went and write a final project report. It is important to learn from past successes and failures. While the project is still fresh in your mind, analyze what went well and generate ideas for how the project could have been improved. Get inputs for these ideas from the project team members. Write a report which shows how the project contributed to the organization's strategic and financial plans, highlights the accomplishments of the project, and summarizes the ideas for improvement. Make sure to give credit where credit is due. Recognize the support from the various functional areas. This report is a record of what you and your project team accomplished. It can also be used by other project managers as a learning document.

Follow through on your commitments about recognition and employee development. As part of setting performance expectations for your team, you should have outlined the positive consequences upon successful completion of the project. This may have included group awards or recognition, recommendation for promotions, or desirable follow-on job assignments. Whatever commitments you made to the project team, make sure you live up to these commitments if the performance criteria is met.

Transfer the technology. Before any project is completed, the information developed during the project needs to be transferred to the appropriate personnel. The recipients are your customers. The technology transfer is not complete until they are satisfied.

Transfer team members to other projects. Long before the project ends, make sure there is a plan for each project team member's next assignment after completion of your project. This plan should include the timing of when the transfers will take place based on milestones. Once those milestones are met, follow through on your commitment to transfer the employees.

Recognize the contribution of others. No project can be completed by the project manager alone. Successful completion of the project objectives requires the continual support and commitment of many other people including the project team, the functional managers, and the executives. Recognize the contribution of others, and express appreciation for their support. Try to do this in a public way. Possible ways are writing letters to them and their managers, writing an article in the company newsletter, having them meet with satisfied customers of the project, and recognizing them at the project completion meeting. Recognizing the help you receive from others goes a long way towards maintaining constructive relationships and will help you win their support for future projects.

Hold a project completion meeting. This is the final status review meeting. Invite all people involved with the project and their managers. Briefly cover items from your final project report: how your project contributes to the organization's strategic and financial plans, project accomplishments compared with the initial project objectives, and ideas for improving future projects. Recognize prominently the contributions made by others.

Enjoy yourself at the project completion meeting. Managing a project is a long, arduous journey. There are many twists and turns throughout the project, and it requires the concentrated and sustained effort by you and your team. This is a time to pat yourselves on the back. Feel good about what you have accomplished.

Chapter 26

CONCLUDING REMARKS ABOUT PROJECT MANAGEMENT

One of the biggest challenges you face as an engineer is managing projects. It requires you to balance technical, economic, and human factors.

Keep in mind your most critical responsibility as a project manager is planning. Utilize as much information as possible to develop the plan. This is necessary to develop sound estimates for schedules, resources, and costs.

While implementing the plan, continually monitor progress. When problems arise, take prompt corrective action. In this way you minimize the impact the problems have on the schedule and costs, and you show your team you are willing to take an active role in resolving conflicts.

Finally, regularly provide meaningful project status reports to the executives. This minimizes the need for executives to be involved with all the day-to-day details of the project. The executives will be knowledgeable about the project and can help support you when you need their help to resolve conflicts or clarify priorities. Also, meaningful project reports keep you visible as the project leader. Performing well as a project manager will have a tremendous impact on the success of your engineering career.

KEY POINTS: Project Management

1) Projects need to tie in with the strategic plan for the organization. Therefore continually tie the project's importance back to the big picture for the organization. This leads to greater commitment from executives, functional managers, and project team members.

2) Projects must be evaluated in terms of financial benefits and costs, risks, and other factors important to the organization. See how the project impacts the key financial measures for the organization.

3) A key to successful project management is developing a thorough plan. This plan should include the following elements:

   - project objectives
   - work breakdown structure
   - list of team members
   - responsibility matrix
   - schedule
   - cost
   - project closure

4) Another key for successful project management is practicing effective communication skills. Set clear performance expectations, win support from functional managers, focus on team building, resolve conflicts, run effective meetings, write meaningful project status reports, and take corrective action when necessary.

5) Close the project on a strong note. Analyze the project, and write a final project report. This is a record of what you and your project team accomplished and can be used as a learning document by other project managers. Follow through on transferring the technology and transferring team members to other projects. Above all, recognize the contributions of others.

SUGGESTED ACTIVITIES: Project Management

1) Run an upcoming meeting as outlined in Table 23.2. Prepare participants before the meeting, be punctual, encourage two-way communication, keep it focused, and summarize main points.

2) Think of a conflict you are currently facing at work, home, or school. Determine the most appropriate conflict resolution mode (shown in Figure 23.1) to solve the problem. Use the collaborative approach, if possible, to resolve the conflict as outlined in Table 23.3. If the conflict involves problems with your boss or co-workers, utilize the ideas presented in Table 3.3.

3) Calculate the net present value (NPV) for a project which has an initial investment of $100,000. The net benefit of this project is estimated to be $40,000 after the first year, $40,000 after the second year, $20,000 after the third year, and $20,000 after the fourth year. Use a discount rate of 10%. From a net present value standpoint, is this project worthwhile to pursue?

    (Answer: $NPV = -\$1892$; Project not worthwhile because the NPV is less than zero.)

4) Calculate the return on investment (ROI) for the project described in Activity (3).

    (Answer: $ROI = 8.99\%$)

5) Plan a project you are working on, whether it's at work, school, or home, by considering the elements of a project plan shown in Table 22.1. Utilize project management software, if possible, to plot both a Gantt chart and a PERT chart for the project. Decide which chart is more pertinent for your purposes.

6) For a project you are managing, work with the functional managers and team members as outlined in Tables 22.5 and 22.6 to gain their support.

7) Develop a consistent format to report the status of your project which is clear and concise. Utilize the examples shown in Figures 24.1 through 24.3 and include graphics whenever possible.

REFERENCES: Project Management

Management Skills

Badawy, M. K., *Developing Managerial Skills in Engineers and Scientists, 2nd Edition*, New York: Van Nostrand Reinhold, 1995.

O'Connor, P. D. T., *The Practice of Engineering Management: A New Approach*, New York: John Wiley & Sons, 1994.

Whetten, D. A., and Cameron, K. S., *Developing Management Skills*, Glenview, Illinois: Scott, Foresman and Company, 1984.

Engineering Economy

Au, T., *Engineering Economics for Capital Investment Analysis*, Englewood Cliffs, N.J.: Prentice-Hall, 1991.

Cassimatis, P., *A Concise Introduction to Engineering Economics*, Boston: Unwin Hyman, 1988.

Fleischer, G. A., *Introduction to Engineering Economy*, Boston: P. W. S. Publishers, 1994.

Grant, E. L., Ireson, W. G., and Leavenworth, R. S., *Principles of Engineering Economy, 6th ed.*, New York: John Wiley & Sons, 1976.

Kleinfeld, I. H., *Engineering Economic Analysis*, New York: Van Nostrand Reinhold, 1993.

Kurtz, M., *Calculating for Engineering Economic Analysis*, New York: McGraw-Hill, 1994.

Lindeburg, M. R., *Engineering Economic Analysis: An Introduction*, Belmont, CA: Professional Publications, 1993.

Thuesen, G. J. and Fabrycky, J. J., *Engineering Economy, 8th Edition*, Englewood Cliffs, N.J.: Prentice-Hall, 1992.

Young, D., *Modern Engineering Economy*, New York: John Wiley & Sons, 1993.

Managing Projects

Ahuja, H. N., *Project Management: Techniques in Planning and Controlling Construction Projects*, New York: John Wiley & Sons, 1984.

Badiru, A. B. and Pulat, P. S., *Comprehensive Project Management: Integrating Optimization Models, Management Principles, and Computers*, Englewood Cliffs, N.J.: Prentice-Hall, 1995.

Badiru, A. B., *Managing Industrial Development Projects: A Project Management Approach*, New York: Van Nostrand Reinhold, 1993.

Badiru, A. B., *Project Management in Manufacturing and High Technology Operations*, New York: John Wiley & Sons, 1988.

Bent, J. A. and Thumann, A., *Project Management for Engineering and Construction, 2nd Edition*, Englewood Cliffs, N.J.: Prentice-Hall, 1994.

Cotterell, M. and Hughes, B., *Software Project Management*, London: International Thomson Computer Press, 1995.

Dabbah, R., *Total Project Management: Strategies and Tactics for the Healthcare Industries*, Buffalo Grove: Interpharm Press, 1993.

DeMarco, T., *Controlling Software Projects*, Englewood Cliffs, N.J.: Prentice-Hall, 1982.

Hajek, V. G., *Management of Engineering Projects, 2nd Edition*, New York: McGraw-Hill, 1977.

Jassim, H. and Craig, S., *People and Project Management for IT*, New York: McGraw-Hill, 1995.

Kerzner, H., *Project Management: A Systems Approach to Planning, Scheduling, and Controlling, 5th ed.*, New York: Van Nostrand Reinhold, 1995.

Leavitt, J. S., *Total Quality Through Project Management*, New York: McGraw-Hill, 1993.

Levy, S. M., *Project Management in Construction*, New York: McGraw-Hill, 1994.

Lewis, J. P., *Fundamentals of Project Management*, New York: AMACOM, 1995.

Ludwig, E. E., *Applied Project Engineering and Management for the Process Industries, 2nd Edition*, Houston: Gulf Publishing, 1988.

Meyer, C., *Fast Cycle Time: How to Align Purpose, Strategy, and Structure for Speed*, New York: The Free Press, 1993.

Naylor, H., *Construction Project Management: Planning and Scheduling*, New York: Delmar Publishers, 1995.

Oberlender, G., *Project Management for Engineers*, New York: McGraw-Hill, 1993.

Ritz, G. J., *Total Construction Project Management*, N.Y.: McGraw-Hill, 1993.

Rosenau, M. D., *Project Management for Engineers*, New York: Van Nostrand Reinhold, 1984.

Shtub, A., Bard, J. F., and Globerson, S., *Project Management: Engineering, Technology, and Implementation*, Englewood Cliffs, N.J.: Prentice-Hall, 1994.

Stasiowski, F. A., *Total Quality Project Management for the Design Firm: How to Improve Quality, Increase Sales, and Reduce Costs*, New York: John Wiley & Sons, 1993.

Turtle, Q. C., *Implementing Concurrent Project Management*, Englewood Cliffs, N.J.: Prentice-Hall, 1994.

Ward, S. A., *Cost Engineering for Effective Project Control*, New York: John Wiley & Sons, 1991.

Wysocki, R., *Effective Project Management: How to Plan, Manage, and Deliver Projects on Time*, New York: John Wiley & Sons, 1995.

CONCLUSION

The purpose of this book has been to show how you can succeed as an engineer whether you are approaching your first job or have been in the profession for many years. Success as an engineer requires developing a portfolio of skills that you can apply. Balance technical mastery with business, quality, and human factors. The book presented principles to give you greater understanding of how your organization functions as a system and how you can make a difference. It showed how you can contribute to the organization's success and your personal success at the same time.

1) To succeed as an engineer, start with yourself. No matter your work situation, you can always start by communicating with yourself. Develop your personal strategic plan. In order to be successful, you need to know yourself and what you want to accomplish. Define what you want to accomplish by the end of your life, considering goals both in your profession and outside your work. Analyze your capabilities and environment. This helps define a plan to develop the capabilities necessary to achieve your goals.

Once you have developed your personal strategic plan, manage your time on a daily basis to focus on your most important tasks. Spend a portion of your time every day working toward accomplishing goals in your personal strategic plan.

2) Next understand the strategic plan for your organization. Understand your organization's goals, capabilities, and environment. The strategic plan identifies the most important tasks for the organization to achieve long-term success. This

plan should be clear and make sense to you. If it does not make sense, question it. Seek clarification.

The ideal situation is when your personal strategic plan aligns with the organization's strategic plan. Tasks that are important for the organization to succeed give you an opportunity to achieve goals in your personal strategic plan. Such an alignment gives you a tremendous amount of motivation and energy.

When there isn't close alignment between your personal strategic plan and the organization's strategic plan, you need to decide how to proceed. Either modify your personal goals or modify your environment to one that gives you better opportunity to carry out your strategic plan. This may mean changing jobs or organizations. In any event, developing your personal strategic plan and understanding your organization's strategic plan is an important step towards your engineering success.

3) Understand how the business runs. Finance is the measurement system to determine how the business is doing. Know the key financial measures for your organization and how you can have a positive impact on them. Use financial linkage analysis to show how your efforts improve the sustainable rate of growth for the business.

Remember to focus on maximizing outputs in high growth environments. This can be done by developing new products, improving the product development cycle, and increasing production capacity. For low growth environments, focus on minimizing inputs such as by lowering production costs.

4) Utilize statistics whenever possible to help improve the product development cycle and to increase manufacturing efficiency. The engineering task is a series of solving problems and making technical decisions. Using statistics enables you to recognize and solve problems faster and to improve the quality of your decisions.

Use Statistical Process Control (SPC) in a production environment. SPC identifies when significant change has occurred. Take immediate action to find causes of significant changes. Also development engineers need to consider process capability when setting specifications for new products and processes.

Use Design of Experiments (DOE) to solve problems in development and production. Matrix designs for multi-variable experiments are superior to the one-variable-at-a-time approach.

Take the time to plan carefully your use of statistics. For SPC, think about how to set up your control charts in order to look at the most important parameters. Set up rational subgroups. For DOE, define the experiment objective, consider all variables, and calculate process variances in order to reach correct decisions based on the data.

5) Take the initiative to establish good working relationships with your manager and peers. Build relationships based on trust. Emphasize face-to-face communications. Good interpersonal communications rely on understanding the viewpoint of others. Use listening skills to understand the other person's frame of reference. Set up feedback systems with your internal and external customers. Handle problems with your manager and peers directly, and work together to develop solutions.

To communicate your points most effectively during face-to-face opportunities, plan ahead. Define your purpose and main points. Profile your audience. Anticipate questions or concerns that may come up. End your discussions by summarizing the key points, and define a specific follow-up plan.

6) When called upon to manage projects, use project management skills to achieve objectives important to the organization. Understand how your project contributes to the organization's success, and continually communicate this link as you request support from executives, functional managers, and team members.

Develop a well thought out project plan. Define clearly the project objectives. Work closely with the functional managers to develop a work breakdown structure and a list of project team members. Then work closely with the functional managers and team members to set the responsibility matrix, schedule, cost, and plan for project closure. Use historical data as a guide whenever possible to help determine schedule and cost.

Planning is an iterative process. If the schedule or cost projections do not meet the project objectives, work with the project team to develop solutions for schedule or cost improvements.

Continue to practice strong communications skills. Work closely with the functional managers and your project team to build constructive relationships and win their support. Remember to recognize the contributions of others. Set clear performance expectations. Keep everyone informed by issuing regular status reports and holding project status meetings. When conflicts arise, take quick corrective action to resolve them.

Managing projects is one of the biggest challenges you encounter as an engineer. Not only are there technical challenges, but there are also scheduling, budgeting, and interpersonal communication challenges. You are very visible as a project manager. Practicing the skills presented will help you be successful.

Besides all the skills mentioned, engineers also need to focus on an area of technical expertise. It is impossible to succeed as an engineer without continually developing your technical skills. Read technical journals, attend seminars and conferences, take college courses, and develop a network of professionals in your field to stay up-to-date.

The engineering profession is becoming increasingly complex and challenging. The technical challenges are growing dramatically. The business environment is becoming more competitive on a global scale. Both these factors place greater pressure on engineers.

In this environment, you need to view your profession through a new paradigm in order to maximize your personal success and maximize your benefit to your organization. You need not only excellent technical knowledge and skills, but you also need to apply a broader set of tools to make a tangible, positive difference.

To do this, know how the business is run. Understand the strategic plan and key financial measures and how you can impact them. Utilize statistics to improve the quality of your decisions and to speed up the development cycle. Communicate effectively with others to build constructive relationships and to achieve project objectives that are important to your organization.

To succeed as an engineer, start with yourself. Understanding the principles covered in this book is an important first step. Applying these skills is the next step.

I welcome your comments. Please let me know the ideas from this book that work particularly well for you. I am also interested in other ideas you have. Contact me through:

J & K Publishing
P. O. Box 87204
Vancouver, WA 98687

thy@pacifier.com

INDEX

double-sided hypothesis 219-220
DRAMs 69-70

E

earnings. *See* retained earnings
earnings retention rate 101, 106, 126,
    133-134
engineering economics 295-301
    cash flow 295-297
    equivalency 297
    future value 297-298
    net present value 298-300
    present value 297-298
    return on investment 300-301
engineer's responsibilities
    competitors, knowledge of 89
    customer needs 88-89
    finance, understanding 98, 167, 354
    finances, making an impact on
        149-157, 167, 354
    financial information, getting 150-151
    product development 89-90, 153, 167
    productivity, improving 151-157, 167
    project management 288, 349,
        355-356
    relationships, establishing 25-26,
        45-49, 355
    statistics, utilizing 173, 284, 354
    strategic plan, understanding 74-75,
        92, 353-354
    strategic planning, business 61-62,
        88-90, 92-93
    strategic planning, personal 20-24, 52,
        353
    technical expertise, developing 356
    time management 24-28, 52, 353
equity 106-107, 110, 112, 118-120
exchange rates, currency 158-166, 167
    effect on international competitiveness
        163-166, 167
    factors influencing 160-163, 167
    purchasing power parity condition
        160-163
executive review meetings 332-333
executive's responsibilities
    financial goals 149-150
    priorities, setting 341

project evaluation 293, 301-303, 329
project plan, approving 304, 328, 329
role for projects 311
strategic planning, business 61-62, 72,
    92, 293-295
experimental error 174, 181-183
    alpha 220-222
    beta 220-222
experimental space 218
experiments, full factorial 251-256
    effects, determining 251-252
    example 253-256
    trials, displaying 251-252
experiments, multi-variable 248-283,
    284
    full factorial 251-256
    general principles 248-256
    interactions 248-250
    interactions, estimating 260
    main effects, estimating 260
    matrix design summary 258
    matrix designs 256-283
    one-variable-at-a-time 250-251, 284
    resolution III 259
    resolution IV 257-259
    resolution V 257
    steps for designing 257-259
    trials, displaying 251-252
    variance, estimating 260-261
experiments, one-variable-at-a-time
    250-251, 284
experiments, single variable 227-248,
    284
    comparison of paired data 244-247
    definition of cases 227, 230
    estimate of variance known 242-244
    example, single population and
        estimate of variance known
        243-244
    example, single population and
        variance known 230-231
    example, single population and
        variance unknown 236-237
    example, two populations and
        variances equal and unknown
        238-240

ORDER FORM

 Telephone orders: Call (360) 253-9532.
Have your Visa or MasterCard ready.

 Fax orders: (360) 253-4084 **(NEW)**

 E-mail orders: thy@pacifier.com

 Mail orders: J & K Publishing
P.O. Box 87204
Vancouver, WA 98687

Name: _____

Company name: _____

Address: _____

Address: _____

City: _____ State: _____ Zip: _____

Number of copies of "How to Succeed as an Engineer"

_____ × $29.95 = _____

Shipping:
   Priority Mail: $3.50 for the first book and
      $2.00 for each additional book   _____

Subtotal   _____

Sales tax: Add 7.6% for books shipped to Washington
   addresses   _____

Total   _____

Payment:
☐ Check or Money Order

Credit card: ☐ Visa   ☐ MasterCard

Card number: _____

Name on card: _____ Exp. date: ____ / _____